Henry Erni

Coal, Oil and Petroleum

Their Origin, History, Geology, And Chemistry

Henry Erni

Coal, Oil and Petroleum
Their Origin, History, Geology, And Chemistry

ISBN/EAN: 9783744758888

Printed in Europe, USA, Canada, Australia, Japan

Cover: Foto ©berggeist007 / pixelio.de

More available books at **www.hansebooks.com**

COAL OIL AND PETROLEUM:

THEIR ORIGIN, HISTORY, GEOLOGY, AND CHEMISTRY,

WITH A VIEW OF THEIR

IMPORTANCE IN THEIR BEARING UPON NATIONAL INDUSTRY.

BY

HENRI ERNI, A. M., M. D.,

CHIEF CHEMIST TO THE DEPARTMENT OF AGRICULTURE;
FORMERLY PROFESSOR OF NATURAL SCIENCE, UNIVERSITY OF VERMONT;
AND LATELY PROFESSOR OF CHEMISTRY AND PHARMACY, SHELBY
MEDICAL COLLEGE, NASHVILLE, TENN.

PHILADELPHIA:
HENRY CAREY BAIRD,
INDUSTRIAL PUBLISHER,
406 Walnut Street.
1865.

Entered according to Act of Congress, in the year 1865, by
HENRY CAREY BAIRD,
in the Clerk's Office of the District Court of the United States
in and for the Eastern District of Pennsylvania.

PHILADELPHIA:
COLLINS, PRINTER.

PREFACE.

THE main portions of the present treatise on coal oil and petroleum were written for and published by the *Sunday Morning Chronicle* of this city, several months ago. The number and length of these articles being beforehand limited, I was compelled, for the interest of the readers, to convey as much useful information as I could, sometimes at the expense of clearness and a more proper arrangement.

Petroleum being a subject of most recent study, no books have as yet appeared which presented a full and comprehensive statement of even the most necessary and best established truths. The facts of this treatise had to be collected and sifted with a good deal of labor from many foreign and home journals to which the author had access.

The articles anonymously contributed by me to the *Chronicle* having excited somewhat of interest, and having been favorably noticed, I yielded to the desires of disinterested friends, and now bring them again before the public in a more extended form. I do it, however, with timidity, for to rewrite and rearrange the entire bulk of material was incompatible with my regular duties and feeble health. All that I could do since the first writing, was to add and improve here and there, in attempting to bring the work up to the most recent date.

On the manufacture of kerosene oils,

the following books may be consulted to advantage:—

Dr. Thomas Antisell. Manufacture of Photogenic Oils from Coal, &c.

Abraham Gesner. A Practical Treatise on Coal, Petroleum, and other Distilled Oils. New York.

Dr. Theodore Oppler. Handbuch der Fabrikation Mineralischer Oele, &c. Berlin, 1862.

The most valuable treatise on petroleum, with which I am acquainted, is that of *Tate*, published in London, which I was unable to obtain, however, until a few days ago, when it was kindly forwarded to me by the publisher of this book.

HENRI ERNI.

WASHINGTON, D. C., April, 1865.

CONTENTS.

CHAPTER I.
	PAGE
SOME FACTS IN REGARD TO SCIENTIFIC DISCOVERIES	13

CHAPTER II.
DRY OR DESTRUCTIVE DISTILLATION OF ORGANIC BODIES. PRODUCTS OBTAINED FROM WOOD, COAL, OR TURF	24
Ingredients of Beech-wood Tar	31

CHAPTER III.
PRODUCTS OF THE DISTILLATION OF CANNEL COAL AND THEIR CHEMICAL COMPOSITION	40

CHAPTER IV.
MANUFACTURE OF PHOTOGENIC OILS AND OTHER USEFUL PRODUCTS FROM COAL, WOOD, AND TURF—VARIATION OF THE RESULT ACCORDING TO THE TEMPERATURE EMPLOYED IN DISTILLING	44
Distillation of Coal Tar	45

CHAPTER V.
PURIFICATION OF COAL OIL OR KEROSENE, AND OF BITUMINOUS OILS, TOGETHER WITH A BRIEF HISTORY OF THESE OILS; AND COMPARISON OF ARTIFICIAL PRODUCTS WITH THOSE FOUND IN NATURE	60
Brief History of Bituminous, and Kerosene or Empyreumatic Oils	62
Comparison of Artificial Products with those found in Nature	66

CHAPTER VI.
PETROLEUM OR ROCK OIL—ITS CHEMICAL COMPOSITION—ILLUMINATING POWER	70

CONTENTS.

CHAPTER VII.

	PAGE
REFINING OF PETROLEUM	84
Illuminating Power of Petroleum	87

CHAPTER VIII.

HISTORY OF PETROLEUM OR ROCK OIL	89

CHAPTER IX.

BORING OF OIL WELLS	109

CHAPTER X.

ORIGIN OF PETROLEUM	124
General View of the Geological Distribution of Petroleum in the United States and Canada	155

CHAPTER XI.

PREPARATION OF ANILINE DIRECTLY FROM COAL TAR; AND ITS PROBABLE ORIGIN—ARTIFICIAL PREPARATION OF ANILINE FROM BENZOLE; TRANSFORMATION OF THE LATTER INTO ANILINE—PROPERTIES OF ANILINE—CHEMICAL TEST FOR BENZOLE—COLORING PRINCIPLES DERIVED FROM ANILINE—THEIR MODE OF PREPARATION AND APPLICATION IN DYEING 160

Artificial Preparation of Aniline 165
Preparation of Aniline Colors 170

APPENDIX.

AMOUNT OF PETROLEUM EXPORTED FROM NEW YORK IN 1863 AND 1864, AND THE COUNTRIES AND PLACES TO WHICH IT WAS SENT 182

AVERAGE PRICES OF PETROLEUM IN 1864 AT NEW YORK AND PHILADELPHIA 186

COAL OIL AND PETROLEUM.

CHAPTER I.

SOME FACTS IN REGARD TO SCIENTIFIC DISCOVERIES.

No man, even of the most ordinary pretensions, dare any longer remain blind to the incalculable influence which natural science, and more especially the branch of chemistry, exerts on all the mechanic arts; nor to the great economical benefits which the practical application of its principles secures to all kinds of business. The hidden forces of nature are made daily more subservient to the commands of the human intellect; and with the constantly increasing acquisition of true knowledge, mankind reaps more and more from bounteous nature the riches of her stores in agriculture, mining, art, and manufacture; securing wealth, comforts, and happiness never before enjoyed.

Without earnestly contemplating the vast advantages in the possession of the race, we can scarcely comprehend how much this age of steam and electricity and science in general differs from all preceding ones. Alluding, for example, to the immense national importance of coal as a fuel, Professor Hitchcock says:—

"It is ascertained that, by the same process of growth and decay, beds of coal have accumulated in the United States over an area of more than 200,000 square miles, and probably many more remain to be discovered. Yet, upon a moderate calculation, those already known contain more than 1,100 cubic miles of coal, one mile of which, at the rate it is now used, would furnish the country with coal for one thousand years, so that a million of years will not exhaust our supply. What an incalculable increase of the use of steam, and a consequent increase of population and general prosperity, does such a treasure of fuel open before this country!"*

* "Professor H. D. Rogers, in his *Geology of Pennsylvania*, shows that a vein of coal four feet thick, yielding one yard net of coal, will produce 5,000 tons of coal per acre, which coal possesses a power equal to the life-labor of more than 1,600 men. A square mile of such a vein contains 3,000,000 tons—equal to the life-labor of 1,000,000 men. 'In coal, this (the labor-power of a man for his life),' says an English reviewer of Professor Rogers's volumes, 'is represented by three tons; so that a man may stand at his door while this quantity of coal is being delivered, and say to himself: There, in that wagon, lies the mineral representative of my whole working life's strength.' When we contemplate the further and indisputable fact that one man can, unaided, and under disadvantageous circumstances, mine in ten hours this quantity of coal, we need not be surprised that Peter Barlow, the distinguished engineer, after a full examination of this subject in all its phases, should have said:—

"'It seems, indeed, a reasonable inference from all that has now been stated, that man was designed by his Maker for a higher principle of action—for the exercise of skill, and for invention; to regulate the action of the lower animals to the purpose of labor; to convert air, fire, and water to his service, and only where skill and direction are required, to become himself a mechanical agent.'"— Baird. *Protection of Home Labor.*

From the long list of most important scientific discoveries and inventions, we will merely mention a few: as the electric telegraph, the art of photographing, electrotyping, the application of ether and chloroform to allay human suffering, the manufacture of gun-cotton and collodion; of artificial mineral gems, increasing so that the hopes of chemists have even reached to the producing diamonds or crystallized carbon itself; Dr. Gall's improvements in the manufacture of wine; the artificial preparation of mineral waters, sparkling wines, and of fruit essences, such as the pear, apple, &c., with which the English bon-bons in the market are flavored; the illumination of our houses with gas; the manufacture of ultramarine blue (once confined to the precious lapis lazuli) out of Glauber salt, clay, and charcoal. The manufacture of friction matches out of phosphorus is also worthy of mention. The use of ordinary phosphorus being dangerous, on account of its poisonous property and easy inflammability, chemistry soon showed how to prepare it in a masked or allotropic form, at once perfectly harmless when swallowed in large doses, and safe to be carried, as it inflames only at 500 degrees Fahrenheit. The fumes of ordinary phosphorus in lucifer match factories frequently gave rise to necrosis, a dangerous disease of the upper jaw bones. The new red form of phosphorus gives off no fumes. Still later, friction matches are made without any phosphorus at all.

Metals of the utmost value, in a technical point of view, have been isolated from material among the most abundant on our earth's surface. Aluminum is employed by jewellers as a substitute for silver. Boron crystals have been successfully used in watches instead of jewels. Magnesium has the property, not only of burning like steel wire in oxygen, but in the open air, and with a light so intense that it can be seen twenty miles at sea. It may be inflamed in a candle, and thus light up a room, a cavern, or an ancient pyramid with wonderful brilliancy. It has been ascertained that a wire of 0.297 millimetre in diameter (a millimetre being 0.393 of an inch) will give as much light as twenty-four good stearine candles of five to the pound. Magnesium wire is now manufactured on a large scale by Sonstadt, in Germany. Magnesium light being chemically active, it furnishes to the photographer a substitute at night for sunlight, a good negative picture being obtained in fifty seconds.

Some of the most recent discoveries prove in a remarkable degree how far science can supply all reasonable demands of practical industry. Considering the enormous consumption of paper in meeting the wants of daily literature and newspapers; the steadily increasing use of tapestry, pasteboard, and papier-maché, packing, and the consequences of general waste, the question "How shall the future production keep pace with the demand?" was at one time becoming an alarming one. The paper manufacturers met daily with greater difficulties in procuring raw material. In spite of every stimulus given to the profession of rag picking, the supply was necessarily a limited one. How, then, supply the place of linen rags?

The scientific reasoning amounted simply to this: All kinds of paper are made up of woody fibre (*cellulose*); the rags are but utilized vegetable fibres derived from flax, hemp, cotton, &c. Could similar material, although perhaps in a different form, be obtained, success was certain. In looking for such a substance, common straw suggested itself; this was worked, together with rags, into paper pulp, and the industrial riddle was solved, as far, at least, as the coarser kinds of paper used for packing, &c. were concerned.

As early as 1772, Dr. Schaeffer, of Regensburg, published experiments on the manufacture of paper from different materials. Indeed, his directions were printed upon paper made from Indian corn-husks. This discovery was, however, unheeded, and the process of manipulation lost. In 1856, Moritz Diamant, a writing teacher, from Bohemia, again directed the attention of the Imperial Government of Austria to maize straw. The Government at once took up the matter, and had lately, at its own expense, a large paper mill established, which is now in successful operation. All kinds of writing, printing, drawing, tracing, and colored papers are made of almost matchless beauty. Indeed, it was ascertained by further experimenting and perfecting, that corn-husks may, by means of proper machinery, be spun and woven into cloth suitable for bags, rough towels, oil-cloth, &c.

A large number of beautiful specimens of paper and other fabrics were presented by the Imperial Government to the Commissioner of Agriculture, by whom they are now exhibited in his museum at Washington.

Of all recent occurrences in the scientific world, the discovery and development of the properties of petroleum is perhaps the most worthy of notice. When viewed from a national standpoint, it fairly promises to outrival the gold mines of California, in creating altogether new branches of industry. To the scientific skill and zeal of Reichenbach, we owe the discovery of the principal constituents now prepared chiefly from petroleum, but which he first obtained by the distillation of wood.

These are mainly different oils, creasote, and paraffine, a wax-like body, now frequently exhibited in the shape of beautiful translucent candles. The same products were afterwards more abundantly and cheaply produced by the distillation of coal, bituminous slate, and even turf; and this rich field of usefulness soon became of immense importance over the whole continent of Europe as well as this country.

A thin, volatile liquid, called naphtha or benzine, is obtained from coal oil, furnishing a substitute for turpentine. The proper understanding of the process of combustion shortly led to the construction of suitable lamps for burning some of the oily products, which, in point of cheapness and illuminating power, soon became second to gas only.

At the commencement of this century, the means of lighting the dwelling houses of the masses consisted of poor tallow candles, and dim and dirty oil lamps. What inconceivable benefit, then, must these modern appliances to the generation of light bestow upon the poor working classes, whose labor often begins before day and terminates long after night! Other oils were best adapted for lubricating purposes, and varied from the finest watch oil to lubricators for all kinds of machinery. An oily alkaline substance, called aniline, may be extracted from coal oil, which acted upon by chemicals produces a series of rich and brilliant dyes. At this stage in discovery, *native* petroleum was announced, flowing in some localities almost literally like rivers, and prepared directly in nature's own great distillery. Of course this unlooked-for circumstance rendered the manufacture of coal oil, no matter how cheap and abundant the material, at once unprofitable. The owners of factories turned their establishments into distilleries, where native petroleum was purified. In fact, this revelation has put an end to the whaling business, and there seems to be no limit to the practical usefulness and variety of products obtained therefrom.

Let mankind rejoice when a great truth becomes unfolded and bears its fruit; at the establishment of each step in the eternal succession which leads from barbarism and misery to civilization and happiness. But, let us not forget the sleepless nights, the days of toil, the feverish anxiety, and too often the pinching want of some real discoverer and devotee of science, who, with the most noble aims and most unselfish purpose, has worn his life away in the consummation of this same object which has made the nations glad.

If it can be truly said that the history of almost every great discovery has also been the history of suffering, let us heartily wish for that happy millennium to come, when the universal sentiment shall be, *fiat justitia, ruat cœlum.*

CHAPTER II.

DRY OR DESTRUCTIVE DISTILLATION OF ORGANIC BODIES. PRODUCTS OBTAINED FROM WOOD, COAL, OR TURF.

In order to be able to appreciate the full importance of the at present all absorbing theme —petroleum—its origin, its manifold products, and their significance in relation to our commerce, art, and manufactures—it will be necessary to follow its history somewhat in detail. A brief review of the artificial means hitherto employed in obtaining this invaluable article and its sub-products will prepare the way for a more perfect understanding of the principles which govern its accumulation in nature's great laboratory.

Photogenic oils, and the numerous other products introduced gradually into practical life, and now assuming such general notoriety, were first prepared by a "dry" or "destructive" distillation of vegetable matter, such as wood, rosin, &c. We will endeavor briefly to elucidate the principles involved in this chemical process. If a small chip of wood or straw is burned in atmospheric air—or, better still, in pure oxygen gas—its whole organic structure gradually disappears, and nothing remains but some traces of fixed incombustible mineral ingredients called ashes. The elementary components of wood—*i. e.*, carbon, hydrogen, and oxygen—have passed off in the form of gas or smoke, made up simply of water and carbonic acid. If animal tissues, into whose composition nitrogen largely enters, are submitted to this process of combustion, carbonate of ammonia forms an additional part of the smoke. On the other hand, if vegetable remains—stems, roots, mosses, &c.—are heated in close vessels or retorts, whence the atmospheric air is completely shut off, the products formed are very different, more complex in their composition, and much more numerous. Such an operation is called a dry or destructive distillation. The simplest form of this process, and one which has been practised for the last two thousand years, is exhibited in the common charcoal kiln, which has for its sole object the furnishing of *charcoal*, while the other volatile ingredients, now-a-days considered as even more valuable, are suffered to pass into the air. If hard wood, such as beech, is subjected to dry distillation in a retort, and the volatile products are condensed in a suitable vessel or receiver, four principal classes of bodies are obtained, viz:—

1. Gases.
2. Watery fluid.
3. A dark resinous liquid.
4. Charcoal.

Product No. 1 is a mixture of inflammable gases; the most important of which are: marsh gas $= CH_2$ or C_2H_4. Olefiant gas $= C_2H_2$ or C_4H_4. Hydrogen, carbonic oxide $= CO$. Carbonic acid $= CO_2$, and sulphuretted hydrogen $= SH$. The latter is particularly apt to contaminate coal gas. It is derived from the pyrites contained in the coal.

As early as 1709, Lebon, a French engineer, conceived the idea of turning these carbo-hydrogen gases to the practical use of illumination, and actually lighted his house and garden at Paris in this manner. Murdoch, in England, afterwards substituted coal gas for the same purpose, and made a public exhibition in 1802 by illuminating his residence. Pettenkofer has lately invented an improved apparatus for the manufacture of gas from wood, and has shown that it has many advantages. It is obtained more abundantly, and has a greater illuminating power than coal gas. It is purer, having no sulphurous or ammoniacal odor. In Paris, they have begun to use native petroleum for this purpose, and several patents for the manufacture of gas from this oil have lately been taken out in this country. The process of making it is simple, it requiring no purification, and the apparatus is cheaper and lasts much longer than that for coal gas. It is not improbable that this now so abundant substance, being so cheap a source, not only of illumination but of caloric, may yet in part supersede coal itself. It is an interesting fact that, several years ago, the consumption of gas alone in London had reached the astounding sum of seven billions of cubic feet annually. To make this gas eight hundred thousand tons of coal are required, while the length of the main pipes through the streets of the city amounted to over two thousand miles.

Product No. 2 constitutes an acrid liquid, known to chemists as pyroligneous acid, or wood vinegar; it is much used in the preparation of acetates, such as acetate of iron, of lead, of soda, &c., which are in turn employed in dyeing and calico printing. Again, if pyroligneous acid is slowly redistilled, crude pyroxilic spirit, or wood-alcohol, passes over, a fluid having a disagreeable taste and smell, but which, owing to its cheapness, is largely consumed on the continent, especially in laboratories, and often burnt in lamps instead of alcohol. There is but little doubt that it would answer the conditions requisite for the preservation of anatomical specimens—a significant hint to manufacturers, when we consider the present enormous prices of alcohol, $5 per gallon, owing to which some of our large zoological museums (as that of Prof. Agassiz, in Cambridge) may yet become seriously embarrassed. Its solvent properties closely resemble those of alcohol, all substances soluble in the latter liquid being equally so in wood-alcohol.

Owing to the high duty on spirits of wine in Great Britain, a mixture of alcohol and wood-spirit has long since been brought into use instead of alcohol. It is called methylated spirit, and is unfit for beverages or perfumes, but may be employed in the manufacture of fulminate of mercury (for percussion caps), of chloroform, ether, &c.

Product No. 3 is a wood-tar, a thick liquid, insoluble in water, but soluble in alcohol; it was formerly used principally as a wood-preserver for tarring and calking ships, but later it proved to be an important source of both photogenic and lubricating oils—subjects to be more fully spoken of hereafter.

Product No. 4 is the charcoal remaining in the retort; it is used as an article of fuel, or as a reducing agent in metallurgy.

We have reason to believe that some at least of these secondary products of dry distillation were known to the ancient Egyptians, who are said to have employed crude pyroligneous acid, containing creasote, as a flesh preserver in embalming their dead; according to others, they used native bitumen, or our present petroleum. From a limited experience in embalming, we believe crude light petroleum, or purified commercial oils—in that case charged with some creasote—to be well adapted for preservation, being injected into the veins in the usual manner. Be this as it may, it is only of late that the finer and more subtle qualities of tar, whether native or obtained by the distillation of wood, bituminous coal, or turf, have been brought to light. This hitherto ill-reputed, filthy compound has been shown to contain quite a number of bodies most useful in the arts and manufactures. For this reason it has not only attracted the eyes of the scientific world, but promises to become hereafter an almost boundless source of domestic comfort and happiness.

Ingredients of Beech-wood Tar.

Reichenbach, an Austrian chemist, while operating upon beech-wood tar in the years 1830–35, discovered and isolated the following ingredients:—

Light Oil, or Eupione, from εὖς, good, and πίων, fat, is an inodorous, insipid, limpid, colorless liquid of the specific gravity 0.655; it burns with a brilliant flame, and is miscible with other oils and ether. It boils at 116° F. Its composition is C_5H_6. Frankland considers eupione to consist principally of hydride of amyle. The less volatile portions of the lighter oil contain wood-spirit, acetone, and the hydrocarbons, benzole, toluole, and xylole; these latter may be removed by agitation with sulphuric acid with which they form colligated acids. Reichenbach gave it the name quoted, under the impression that it was a distinct organic substance, and not a mixture of many different liquids, boiling at different temperatures, as has since been shown. Its composition will be noticed when we consider it as one of the coal oils.

Heavy Oil.—This is collected after the eupione has almost ceased to distil over. It is a fatty mixture, containing some of the substances belonging to the light oil, and several oils heavier than water, namely:—

Picamar, from pix and amarus, is a viscid, colorless, oily liquid of an intensely bitter taste. To this principle tar owes its bitterness.

Kapnomore, from καπνὸς, smoke, and μοίρα, part. —This forms a colorless oil, having a taste like ginger, and producing a sense of suffocation. Besides these components, the heavy tar-oils contain creasote and paraffine, about to be described, and some other compounds extracted at a higher temperature by Laurent, such as chrysene $= C_{12}H_4$ and pyrene $= C_{30}H_{12}$.

Creasote.—$C_{16}H_{10}O_2 = HO, C_{16}H_9O$ (?) Its preparation is tedious. The heavier portions of the oil obtained from wood-tar after being washed with a solution of carbonate of soda, are submitted to distillation, by which they are further separated into a portion lighter than water, and into another which sinks in this liquid. This heavier oil is then treated with a solution of potash of specific gravity 1.12. By this means the creasote is dissolved and the greater part of the hydrocarbons which accompanied it are separated. The alkaline solution, after being decanted from the hydrocarbons, is boiled gently in an open basin, with a view to oxidize a portion of the impurities. When cold, dilute sulphuric acid in slight excess is added to the liquid, by which means the creasote is set at liberty. To purify it, it has to be redistilled with water, again treated with concentrated solution of potash, then with dilute sulphuric acid, and then redistilled with water. Finally, the creasote must be digested upon chloride of calcium and distilled by itself. It will then have the boiling point 398° F., and does not become brown by keeping. When pure it is an oily, colorless, neutral fluid, exhibiting a strong, peculiar smoky odor, and sharp, burning taste. Its specific gravity is 1.057; its boiling point 398° F. It is inflammable, soluble in acetic acid, alcohol, ether, and benzole, and coagulates albumen instantly.

It forms a large part of the heavy oil of tar passing over toward the end of distillation; the nauseous smell of tar or of native petroleum, is mainly due to this substance. Creasote is but slightly soluble in water; it has most powerful antiseptic properties. Thus, a piece of flesh steeped in a very dilute solution of it, dries up into a mummy-like substance, which thence refuses to putrefy. Tongues and hams may be almost instantly cured by immersing them in a mixture of one part of creasote, and one hundred parts of water or brine. Dentists employ it for the purpose of relieving toothache arising from decaying teeth. In a very diluted form it is a most valuable application in cases of fetid ulcers, hospital gangrene, and many cutaneous affections, as itch, &c. In the smallest quantities it prevents or stops the fermentation of wine, cider, beer, &c. It is this substance which imparts to wood smoke, produced by incomplete combustion, its qualities of preserving meat. The eyes of many a poor wretch of boarding-house experience testify to its pungent properties, as he sheds compulsory tears over a bad fire. The creasote at present most found in commerce is either prepared direct from coal, or from petroleum during the process of purification, and has a dark color generally. Wood and coal creasote, though distinct bodies, exhibit, in some respects, identical properties; both are of course dangerous poisons. Wood creasote is probably a homologue of coal creasote.

Coal tar creasote, syn. with phenic or carbolic acid, hydrated oxide of phenyl, and phenol, $= C_{12}H_6O_2 = HO,C_{12}H_5O$. Specific gravity $= 1.065$; boils at 369° F. It forms also a product of the distillation of gum benzoine, of the resin of the *Xanthorrhœa hastilis*. Stædeler found it in the urine of the cow. Its solutions do not redden litmus paper. A drop of it let fall upon paper produces a transient greasy stain. If a splinter of deal be dipped into a solution of phenic acid and then into nitric or hydrochloric, the wood as it dries becomes blue. Phenic acid, when heated with ammonia in a sealed tube, is partly converted into water and aniline; with caustic potash it forms a crystalline compound. It is now more particularly employed as a valuable permanent dye-stuff for silk and woollen fabrics. Carbolic acid, when treated with nitric acid at a moderate heat, yields carbozotic or picric acid of a yellow color; by concentrating this liquor by evaporation, we obtain yellow, scaly crystals. Picric acid is intensely bitter, like quinine, and may prove a good remedy for intermittent fever. Ale and beer have been repeatedly found to be adulterated with it. Like all the tar colors, its dyeing qualities when in solution are most intense, *i. e.*, a very small weight of the material goes very far. Silk and woollen goods, without further preparation, when brought into the solution even cold, assume a magnificent yellow color, throwing far into the shade those obtained from other dyes. Cotton fibre affords less attraction for the dye. Picric or trinitro-phenic acid, as it is often called, may, however, be obtained from a great variety of substances, when acted upon by hot nitric acid, such as indigo, aniline, saligenine, salicylous and salicylic acids, salicine, phloridzine, silk, aloes, coumarine, and many gum-resins, etc. Picric acid has the composition $(HO, C_{12}H_2(NO_4)_3O)$. It may be fused to a yellowish oil, and even be partially sublimed, but if suddenly heated it explodes.

Paraffine or *tar-wax* is another useful and interesting body. It comes over toward the last stages, when crude tar is rectified. It is particularly abundant in beech tar, but occurs in the tar of both animal and vegetable substances, and in all kinds of American petroleum. Its specific gravity is 0.870, and it fuses at 110.7° F. It is a pearly white, tasteless, and inodorous solid, miscible when melted, in all proportions, with fixed and volatile oils; the strongest and most corroding acids and alkalies have no effect upon it, whence its name, from *parum, affinis.*

It burns with a bright, white flame, without smoke. It is now much employed as a material for candles, which for matchless purity and lustre are without a rival, the best and most costly of wax tapers not excepted. Paraffine possesses many properties which render it useful in the laboratory. It may be advantageously substituted for oil in baths, as it endures a high temperature without evaporating or emitting any unpleasant odor. Bibulous paper, after being soaked in it, may be kept several weeks in concentrated sulphuric acid without undergoing the slightest alteration. Hence paraffine forms an excellent coating to labels; hydrofluoric acid, even, does not act upon it except it be heated. It appears also to be useful in preserving fruits. Apples, pears, &c., coated with it retain all their freshness during many months. Perhaps it will be proved that flowers might thus be likewise preserved.

Paraffine is now most abundantly obtained by distillation from cannel coal, or native petroleum, when it comes over with certain *isomeric* oils, passing into the receiver toward the close of distillation. In other words, the oily mixture which remains after most of the photogenic oils have passed over, and from which the solid paraffine may be obtained, is called paraffine oil. For experimental purposes on a small scale, paraffine may be most readily prepared by distilling beeswax with lime.

Asphalt, or pitch, is the fixed residue left after distilling tar; like the native asphalt, it is used for varnishes and as an ingredient for making lampblack, which latter is again employed in preparing printers' and lithographers' inks.

Reichenbach, however, did not yet procure any of these substances in quantities sufficient to turn his important discoveries to practical advantage. Mansfield and Young, of practical England, took out patents, in 1848, for preparing, out of coal tar, obtained as a bi-product in gas establishments, paraffine and certain oils suitable for photogenic and lubricating purposes.

CHAPTER III.

PRODUCTS OF THE DISTILLATION OF CANNEL COAL AND THEIR CHEMICAL COMPOSITION.

The products obtainable from coal are still more numerous than those from wood; many of them differ essentially, as would be naturally inferred from the different nature and composition of the material from which they are derived. Wood being rich in oxygen, and poor in nitrogen, furnishes products containing much acetic acid and little ammonia, exhibiting hence an acid reaction. Coal and animal matters, containing, on the contrary, much nitrogen and but little oxygen, yield a good deal of ammonia, imparting to the products an alkaline reaction. Coal tar has of late been shown to contain—

1. Acid oils, soluble in alkalies, such as potash, &c.

2. Alkaline oils, soluble in acids, such as sulphuric.

3. Neutral oils not affected by alkalies and some acids.

No. 1 consists essentially of carbolic acid or coal creasote = $C_{12}H_6O_2$, together with small quantities of the following acids, viz: rosolic, brunolic, acetic, butyric.

No. 2 constitutes but a small bulk of the mass, and consists chiefly of ammonia = NH_3, aniline, syn. with phenylamine = $NC_{12}H_7$, and leucoline, syn. with quinoline = N,C_{18},H_7.

Both of the latter furnish a base for the most beautiful dyes, representing all the colors of the solar spectrum or rainbow, if the yellow shade, obtained from coal creasote, as previously described, is included.

Besides, traces of the following substances have been detected, viz:—

Ethylamine,   Lutidine,
Methylamine, Cumidine,
Picoline,   Pyrrole.
Toluidine.

No. 3, comprising coal oils proper, is composed of a great variety of hydrocarbons, both liquid and solid, the latter being held in solution. All are of different volatilities, viz., boil at different temperatures.

A. Alcohol series of hydrocarbons—
Hydride of Amyle $= C_{10}H_{12}$, boils at 102° F.
Hydride of Caproyle $= C_{12}H_{14}$, boils at 154° F.
Hydride of Œnanthyle $= C_{14}H_{16}$, boils at 208° F.
Hydride of Capryle $= C_{16}H_{18}$, boils at 246° F.

B. Benzole series of hydrocarbons—
Benzole $= C_{12}H_6$, boils at 177° F.
Toluole $= C_{14}H_8$, boils at 280° F.
Xylole $= C_{16}H_{10}$, boils at 263° F.
Cumole $= C_{18}H_{12}$, boils at 299° F.
Cymole $= C_{20}H_{14}$, boils at 341° F.

C. Paraffine series of solid hydrocarbons.

Paraffine = (empirically) $C_n H_n$. Its rational formula is not known, but it appears to be a homologue of olefiant gas $= C_4H_4$. Its specific gravity is 0.870. It fuses at 110.7° F. Its boiling point is upwards of 418° (?).

Naphthaline $= C_{20}H_8$. Specific gravity 1.153; fusing point 174°; boiling point 428°. Beautiful red and blue colors, rivalling those from aniline, have lately been obtained from this pearly-white solid.

Paranaphthaline, syn. with anthracene, = $C_{30}H_{12}$, i. e., it forms, according to Dumas and Laurent, a polymere of naphthaline. Anderson's analysis leads to the formula $C_{28}H_{10}$. Fusing point 416°; boiling point 570°. It forms a white crystalline solid.

Pyrene = $C_{30}H_{12}$.

Chrysene = $C_{12}H_{14}$.

These two solids were obtained by Laurent in the latter stages of the distillation of fatty and of resinous bodies, and in that of coal tar.

CHAPTER IV.

MANUFACTURE OF PHOTOGENIC OILS AND OTHER USEFUL PRODUCTS FROM COAL, WOOD, AND TURF—VARIATION OF THE RESULT ACCORDING TO THE TEMPERATURE EMPLOYED IN DISTILLING.

NAPHTHALINE and paranaphthaline are formed when organic substances are decomposed at a high temperature, as in gas works, where they frequently incrust the pipes leading from the retorts.

Paraffine accompanies the heavier coal oils produced at a more moderate temperature; from this fact it will be perceived at once that the manufacture of illuminating gas and of paraffine oils can never be advantageously conducted at one and the same time.

Distillation of Coal Tar.

In distilling coal tar over a free fire, or by passing steam through the retorts, one of the first products which goes into the receiver is a light, very mobile fluid, known as crude naphtha. Washing, first with dilute sulphuric acid, to remove the basic oils, and next with potassa liquor, to remove any acid oils present, and repeated distillation of the purified mixture, furnishes the so-called rectified naphtha, of which benzole or benzine forms one of the most abundant and useful substances.

Benzole was originally obtained by Faraday from the liquid produced by the compression of oil gas, and named bicarburet of hydrogen. Benzole may be easily procured in small quantities by distilling one part of benzoic acid with three parts of quicklime; the distillate should be agitated with a weak solution of potash, and the benzole which rises to the surface be dried by digestion upon chloride of calcium; after which it may be obtained pure by redistillation. Benzoic acid yields about a third of its weight of benzole.

From coal naphtha—the cheapest and most abundant source of benzole—it may be procured pure by repeated rectifications, and exposing the product to a cold of 32° F., when it solidifies in transparent crystals or camphor-like masses. The other hydrocarbons associated with it remain liquid at that temperature. Two gallons of naphtha furnish thus a pint of pure benzole. Even the commercial, impure, and diluted benzole or coal naphtha constitutes a superior menstruum for oils, resins, and fats, which renders it suitable for family use in removing stains from silks, woollen and cotton fabrics, carpets, &c.* A solution of one part of wax and one of rosin, in two parts of naphtha, forms an excellent furniture polish.

Naphtha, containing no oxygen, may be

---

* A fluid now usually sold in drugshops for this purpose, and labelled benzine, we find to consist merely of the more volatile constituents of petroleum; it answers in a measure the purpose for which it is designed.

advantageously employed for preserving potassium, sodium, manganese, and other oxidizable metals. Care should be taken, however, first to ascertain whether it is truly anhydrous, *i. e.*, deprived of water, otherwise serious explosions might occur.

By its own evaporation it keeps off insects from zoological collections or stuffed animals. Mansfield introduced benzine into the English market more especially as a solvent of caoutchouc and gutta percha. The solution is used in rendering cloth and other fabrics waterproof; also in the manufacture of syringes, surgical instruments, &c.

It replaces oil of turpentine, ether, &c., in preparation of varnishes and paints.

Being very volatile and highly inflammable, it is even more dangerous than turpentine. Hence explosions of lamps and conflagrations in storehouses have been frequent. The law ought to prohibit its storage in the very hearts of our cities, and should regulate its shipping by railroads and vessels, in order to protect human life. Since many of the component parts of coal oil assume the form of vapor at *any temperature above zero*, it is evident that we cannot too carefully guard against those accidents so common to explosive compounds. *Its volatile nature renders it as dangerous as alcohol.* Shall the late terrible catastrophe at Philadelphia pass by unheeded, or will the public, taught by sad experience, at length show their appreciation of these facts?

Naphtha and petroleum exposed to the atmosphere gradually thicken or solidify, for, like many of the essential drying oils, they take up oxygen and form resins or gums. Painters and artists mix their colors with drying oils, such as turpentine, poppy, nut, and flaxseed oils.

Benzole, when treated with concentrated nitric acid, forms nitrobenzole or artificial oil of bitter almonds, used in the art of perfumery.

The following formula will render this process clear:—

$$\underbrace{C_{12}H_6 + NO_5}_{\text{Benzole + Nitric Acid.}} = \underbrace{C_{12}H_5NO_4 + HO.}_{\text{Nitrobenzole + Water.}}$$

By means of nitric acid all the members of the benzole series may be separated from the amyle series of hydrocarbons.

Nitrobenzole or "Essence de Mirbane," as it has been styled, is largely used for scenting fancy soaps, for which purpose, being less affected by alkalies, it is more suitable than the genuine oil. For confectionery shops it is still more preferable, since it never contains, like the other, traces of prussic acid—a fearful poison.

Nitrobenzole may be readily transformed into aniline, to which, in connection with the justly celebrated tar colors, we shall direct the attention of the reader in a separate chapter.

As the distillation goes on and the temperature rises, the heavier and less volatile oils come over, which may, as such, be disposed of to machine shops, or, by redistillation, be changed into light oil and paraffine.

At last the dark-colored paraffine oils appear, which are so much charged with paraffine that, by exposure in open vats, they deposit this body in white scales.

In the still remains a highly carbonaceous residue, or artificial asphalt.

This new branch of industry, arising from the discovery of facts relative to the distillation of vegetable matter, which were hitherto unknown, began to prosper and to develop itself rapidly, especially since it was found that not only coal but such abundant material as peat, and many calcareous schists, would likewise yield oils.

Indeed, science established the fact, that, in the process of carbonization, all vegetable and animal tissues furnish some identical products.

Thus, paraffine results from the preparation of bone black, and also enters into the composition of soot, and even tobacco smoke.

In many countries, mighty layers of bituminous rocks, hitherto barren and unprofitable, could thus be turned into an immense capital. Again, the peat swamps of Ireland constitute a seventh part of the surface of the whole country; peat, as a fuel, situated as it was in the vicinity of the rich English coal mines, was valueless, and, in an agricultural point of view, a curse to the country. Now, the picture was reversed, as thousands of diligent hands, while furnishing human society with most useful products, were producing for themselves their daily bread. And, further still, on the continent of Europe, as in Germany, France, &c., where the production of animal oils and fats is meagre, large tracts of lands were before cultivated with plants yielding oil-bearing seeds, such as rape, flaxseed, camelina, sativa (golden pleasure), &c., upon which wheat and other grains may now be raised.

Tallow and animal oils, employed for candle and soap making, lubricators, &c., may be turned to other account; indeed, chemistry has commenced to convert some animal fats into artificial butter, which, at least, for frying and baking purposes, proves highly valuable. The following results, obtained by chemical analysis, may be taken as the average quantity of products derived from the distillation of schists and turf:—

One hundred parts of a bituminous slate of Wurtemberg yielded—

Tar	9.63
Water and ammonia	8.33
Gas	12.36
Residue (coke, rocky matters)	68.68
	100.00

One hundred parts of the above tar furnished—

Light oil or photagen	24.18
Heavy or lubricating oil	41.94
Paraffine	.12
Creasote	19.04
Carbon residue or asphalt	13.69
Gas and loss	1.13
	100.00

One hundred parts of peat or turf from Hanover, dried in the air, yielded—

Tar	9.06
Ammoniacal liquor	40.00
Coke	35.32
Gas and loss	15.62
	100.00

One hundred parts of this tar gave—

Light oil	19.46
Heavy oil	19.55
Paraffine	3.31
Asphalt	17.19
Creasote and loss . . .	40.49
	100.00

Whence by a simple calculation we find that 100 parts of this turf will furnish—

Light oil	1.76 parts
Heavy oil	1.77 "
Paraffine	0.30 "
Asphalt	1.56 "

The chemical works established in the county of Kildare, Ireland, are capable of working up one hundred tons of peat per day. Every ton of peat yields three pounds of paraffine, two gallons of volatile oil, adapted for burning, and one gallon of fixed oil, for lubricating purposes, all separated from the five to six gallons of tar furnished by one ton of peat. Besides, each ton of peat yields sixty-five gallons of ammoniacal liquor, containing a list of useful substances; for all practical purposes it suffices to mention ammonia, acetic acid, and pyroxilic spirit, &c., already described in connection with wood tar.

One ton of peat affords—
  $5\frac{1}{2}$ lbs. of ammonia,
  5 lbs. of acetic acid, and
  8 lbs. of naphtha.

Besides, some of the tar residue or asphalt may be converted into a valuable grease or lubricator for the axles of carriages, railroad cars, &c. Peat tar itself is an excellent preventive of the fouling of ships' bottoms, successfully resisting those marine incrustations, whether of an animal or vegetable nature, so detrimental and injurious to shipping. An experiment, made in Scotland, proved that one side of a schooner, to which it was applied, presented at the end of six months the same clean appearance as when laid on, while the other side, on which was put the usual composition or paint, became so fouled as to require cleaning during that time.

Variation of the Products of Distillation, according to Different Temperatures.—Experience teaches that the relative proportion and the chemical nature of the products of distillation vary not only with the different materials employed, whether wood, coal, or turf, but also according to the temperature to which one and the same substance is subjected. Thus, if we desire to obtain the greatest amount of permanent gas from coal, it ought to be rapidly decomposed—*i. e.*, heated to a temperature of 800 to 1,000 degrees F. If it be our purpose to procure the greatest amount of fluids, no temperature above 700 degrees F. is admissible. Hence, it follows that the manufacture of illuminating gas, and that of photogenic oils, cannot be combined with economy; neither can the process of tar and coke making with the production of volatile oils, as was at first imagined to be practicable.

Coal heated to a strong red heat, 980 degrees F., yields a maximum quantity of tar, but the fatty bodies separable therefrom by fractional distillation are mainly naphtha (benzole) and naphthaline, a crystallizable solid, but no paraffine. Neither of the first two are desirable constituents of lamp-oils. The chief ingredients of these are compounds evolved after the naphtha ceases to come over, and before naphthaline is produced. That is, to say again what was previously mentioned, the manufacture of photogenic oils terminates where that of illuminating gas commences.

Paraffinized oils are generated from coal between 350 and 700 deg. F. The manufacture of coal oils may be advantageously accomplished at a much lower temperature, by permitting super-heated steam to flow through the retorts, or by conducting the distillation in a partial vacuum, like the boiling down of cane sugar.

The retorts, in which the distillation of coal is conducted, are of iron or clay, and shaped much like those in gas works. The exit tube for the products, inserted at the end opposite to the mouth, is of considerable length, and kept cool by a constant stream of cold water. The gases, after passing through this pipe, enter into a large iron cylinder filled with coke, which removes the last traces of the tar they contain; thence they are suffered to escape into a chimney of strong draft.

The liquid products of distillation flow into a large reservoir, kept at a temperature of 86° F., in which tar separates from the ammoniacal waters; these waters are mixed with the residue of the large retorts, and furnish a rich manure, or are turned into sal-ammoniac.

It has been calculated that in England alone 4000 tons of this salt are annually thus obtained. This important compound was, at one time, imported from Egypt, where it was first prepared by the distillation of camels' dung, near the temple of Jupiter Ammon, from which it takes its name.

The tar and its crude oils are next pumped into a purifying apparatus, made of cast-iron, and mixed with a few per cent. of copperas, to free it from sulphide of ammonium. If the coal contains a good deal of sulphur at first, it is sprinkled over with caustic lime before being distilled, as sulphur compounds impart to burning coal a very bad odor. After this the tar is brought into regular stills, holding several hundred gallons, and heated over a free fire or with superheated steam. The oily volatile products of distillation are condensed in a leaden coil, 30 to 40 feet long.

The following products may be separately collected by a fractional distillation :—*

1. A thin, volatile, very inflammable liquid, lighter than water (spec. grav. 0.8)—*i. e.*, crude naphtha.

2. An oily mixture, heavier than water, called solar oil, which continues to come over until the temperature approaches 400°; it is best suited for a burning fluid for lamps (argands) with a round wick, allowing the air access into the interior of the flame.

3. Paraffinized oil, so called from its containing largely paraffine; it is well-adapted as a lubricator.

Mixtures of 1 and 2 burn very readily, and those of Nos. 2 and 3 furnish an excellent machine oil.

* Dr. Antisell's Treatise on Coal Oils.

The rest of No. 3, not used for mixing, is exposed in vats to a low temperature for several weeks; when it crystallizes it is submitted to the hydraulic press, melted again, and purified with concentrated sulphuric acids and potassa solution.

The residue left in the still forms a tarry mass, which by means of caustic soda may be converted into a black soapy grease used as a lubricator of wagons and railroad cars.

CHAPTER V.

PURIFICATION OF COAL OIL OR KEROSENE, AND OF BITUMINOUS OILS, TOGETHER WITH A BRIEF HISTORY OF THESE OILS; AND COMPARISON OF ARTIFICIAL PRODUCTS WITH THOSE FOUND IN NATURE.

HAVING, on a former occasion, already alluded to the qualities and chemical composition of burning and lubricating oils, showing that these are neutral compounds, being, in other words, indifferent towards acids as well as alkalies, while the pernicious admixtures readily combine with these chemicals, the principle and methods of purification suggest themselves as deserving of a small portion of our attention.

In addition to the redistillation (or rectification) of crude tar oils, whereby traces of tar and other highly carbonaceous solids and liquids suspended in the oil are removed, chemical purifiers are employed; amongst these are principally sulphuric acid and caustic soda. The oil is agitated or churned for several hours with about five per cent. of its weight of sulphuric acid* at a temperature of from 75° to 90° F. It is then allowed to settle, and drawn off into a second purifier, and mixed with five per cent. of caustic soda solution (or lime-water), and the whole stirred for two or three hours, and left to repose; having gone through this process, the oil is once more distilled. Sulphuric acid unites with several heavy hydrocarbons and detaches them from the lighter oils upon which it has no action; the soda answers the double purpose of neutralizing an excess of acid, and of removing creasote or carbolic acid.

* Manganate of potassa and nitric acid are used for the same purpose; others purify each separate oil with six per cent. of sulphuric acid; one-eighth per cent. bichromate of potassa; two and one-half per cent. muriatic acid.

Brief History of Bituminous, and Kerosene or Empyreumatic Oils.

It is on the continent of Europe, where whale and other animal oils and fats are high in price, and the supply of vegetable oils insufficient, that the distillation of natural tar or asphalt and bituminous slate was first resorted to in order to obtain illuminating oils. As early as 1819 the celebrated savant, De Saussure, in Switzerland, distilled bituminous limestone, and pronounced the oil obtained therefrom identical with that derived from the native petroleum of Amiano, in Italy. For more than twenty years past lamp oils were extensively prepared in Germany and France from wood, rosin, schists, and bitumen or asphalt.

Tar oils obtained from animal substances, containing sulphur and phosphorus, have a penetrating offensive odor, and are tedious to purify, and hence less fit for practical purposes.

The manufacture of bituminous oils in Great

Britain and this country is of much more recent growth, because the extensive pursuit of the whale fishery supplied all the wants of the market. Still, the manufacture of volatile oils from coal was first practised in England, and the process was in some respects new; for we must call to mind that it is only quite lately that chemists look upon oils procured from coal, wood, native bitumen, &c., as analogous, if not absolutely identical products. This ignorance must now surprise us still more when we consider the origin of the material and recollect that even before the production of illuminating gas, large amounts of coal were distilled simply to obtain tar to satisfy the necessities of the English navy and mercantile marine—wood tar being, though preferable, too expensive. The English process of obtaining these useful oils from the distillation of coal was, from selfish motives, kept secret at first, and even samples of oils withheld from the exhibitions of all nations at Paris in 1855.

The first empyreumatic oil was manufactured in this country in 1850, at Brooklyn, New York, by distilling wood and rosin together, a jet of high pressure steam being conducted through the retorts. The writer visiting the factory, and being consulted about its utility as a burning fluid, found it would not burn without smoke, even in lamps consuming readily oil of turpentine. There was at that time no lamp invented with a sufficient draft to burn completely such highly carbonaceous oils, whence wood and coal oils, as light-furnishing mediums, found at first but a slow access into dwellings. The earlier constructed lamps separated a very fine carbon or soot, which, settling imperceptibly upon the faces of a company, often presented a ludicrous spectacle. These oils were in earlier times sent to market in a crude and unrefined state, and consequently were ill calculated to win public favor. Their bad creasote-like odor was enough to cause their rejection at first by persons of delicate sensibilities; but gradually every obstacle was overcome, and little more could be objected against their general adoption and use.

The manufacture of coal oil in this country

was first introduced in 1853, and was generally confined to districts where highly bituminous (cannel) coal could be mined at a cheap price. Hence the States of Kentucky, Virginia, Pennsylvania, Ohio, Missouri, and Illinois became great centres for its manufacture.

The Lucesco works in Westmoreland County, Pa., were perhaps the largest in the country, producing 6000 gallons of crude oil a day, which is also rectified there. At Brooklyn N. Y., were located the New York kerosene oil works, producing and refining 1000 gallons of oil daily. This factory was perhaps the only one far removed from the source of material, but New York being the great commercial market and a sea board city, the saving of expenses for transportation of the refined oil, compensated for extra outlays in shipping coal. In Franklin County, Va., near the Kanawha River, was a factory producing 1000 gallons of oil per day. The refining operation was conducted at Maysville, Ky. In 1860 the total number of factories in the United States was over sixty.

Comparison of Artificial Products with Those Found in Nature.

Natural products, closely resembling the artificial ones alluded to, have been found in many localities all over the world for ages past. Some escape as gases from crevices of rocks; others are liquid, and exude through the soil in drops, or spout out through fissures in the rocks like fountains; others, again, of a solid shape, are imbedded beneath the earth's surface. The complete chemical analogy between both classes was not dreamed of for some time, and was only ascertained step by step in the gradual progress of science. Thus the close connection, if not identity, of the natural product, long known under the synonymous names of naphtha, mineral naphtha, petroleum, rock oil, and seneca oil, with coal oil, was experimentally established, its importance appreciated, and its value understood, long after the same article, artificially produced, had attracted, by its vast commercial value, the notice of the civilized world. The following table will enable the general reader to get a comparative view of the two classes of artificial and native products corresponding to one another.

ARTIFICIAL PRODUCTS FROM PIT OR BITUMINOUS COAL.	PRODUCTS OCCURRING NATIVE IN THE EARTH.
1. Illuminating gas.	1. Inflammable gases (sacred fire of the Brahmins), issuing here and there from crevices of the rocks.
2. Thin or light oil of coal tar, containing benzole, &c.	2. Naphtha, a thin and nearly colorless variety of rock oil oozing out of the earth in Italy and Persia, and containing benzole.
3. Thick or heavy oil of coal tar, containing paraffine.	3. Mineral tar, found in many places in Persia, America, and France. It is darker and more viscid than rock oil, and contains paraffine.
4. Artificial asphaltum (pitch of pit coal.)	4. Natural asphaltum (or pitch of Judea), found in the Dead Sea, and other Asiatic seas. It contains paranaphthaline.

ARTIFICIAL PRODUCTS FROM PIT OR BITUMINOUS COAL.	PRODUCTS OCCURRING NATIVE IN THE EARTH.
5. Ammoniacal empyreumatic liquid.	5. Ammonia, issuing in watery vapor, associated with boracic acid, from the earth of Tuscany.
6. Coke as produced and seen in all gas works. It is a porous and light carbon.	6. Anthracite coal in immense beds in Pennsylvania. It is a compact and heavy carbon, owing to the enormous pressure to which it has been exposed.

The following organic substances are allied to the native bitumens mentioned in the previous table.

1. *Elastic bitumen, mineral caoutchouc.*—This curious body has hitherto been found in three places: In a lead mine at Castleton in Derbyshire, at Montrelais in France, and in Massachusetts in America. In the latter localities it occurs in the coal series. It is fusible, and resembles in many respects the other bitumens.

2. *Retinite* or *Retinasphalt.*—It is found in brown coal, and constitutes a fossil resin,

which has a yellow color, is fusible and inflammable, and largely soluble in alcohol.

3. *Hatchetin*, similar to the last named, is met with in mineral coal-beds at Merthyr Tydvil, and near Loch Fyne in Scotland.

4. *Idrialin* is found associated with native cinnabar, and is extracted from the ore by oil of turpentine. It constitutes a white crystalline substance, composed of $C_{42}H_{14}O$; it is generally associated with a hydrocarbon idril, which contains $C_{42}H_{14}$.

5. *Ozokerite*, or fossil wax, occurs in Moldavia and Switzerland in bituminous shale or brown coal. It is brownish and has a pearly lustre. It fuses below 212° F., is easily soluble in turpentine, but with difficulty in alcohol and ether.

CHAPTER VI.

PETROLEUM OR ROCK OIL—ITS CHEMICAL COMPOSITION—ILLUMINATING POWER.

PETROLEUM is named from *petra*, a rock, and *oleum*, oil. This highly important native compound, analogous in every respect to the kerosene oils just described, will next engage our attention. It is, like these, a mixture of a great many chemically different substances, and, as proved by its composition, is evidently of organic origin.

Petroleum and its manifold products find an almost endless application in science, art, and in practical life.

It is used in the preparation of paints and varnishes, the lighter portion or naphtha distilled from it, dissolving caoutchouc, camphor, fatty and resinous bodies generally, and when hot even sulphur and phosphorus. It forms a substitute for fish-oil in tanning. Petroleum soap is already a favorite toilet article.

Aniline and its brilliant colors may perhaps be prepared from the waste petroleum after refining it. Petroleum forms an already valuable caloric.

Thomas Shaw, in an article in the "American Gas Light Journal," in which he advocates the economy of the use of oil as fuel, says, "that the heating value of 100 pounds of coal average quality, spec. grav. 1.279, is equivalent to raising 812,307 pounds of water 1° C. The heating value of 100 pounds of petroleum is equivalent to raising 1,231,600 pounds of water 1° C., making the heating value of coal as compared with petroleum, as 1 is to 1.51. This is the calculated value of their component parts."

It furnishes, as generally known, photogenic and lubricating fluids. It has proved to be an efficient remedy like ordinary coal oil, especially in ulcers and cutaneous diseases. Petroleum vapors are said to act very beneficially in protracted cases of asthma and weak lungs.

Mr. Bobb states that the air in oil pits becomes charged with vapors of an intoxicating effect.

The "American Druggist's Circular and Chemical Gazette," in speaking of a new anæsthetic, says: "Dr. Genges has addressed a note to the French Academy, giving an account of some interesting experiments in trying new agents for diminishing sensibility. He has ascertained that a purified kerosaline, obtained from common petroleum, when vaporized by means of heat, will be found a most valuable anæsthetic."

It is already well understood that amylene gas, closely allied to some of these hydrocarbons, and which is obtained by decomposing chloride of amyle by fused hydrate of potash, acts on the system like chloroform, though with more dangerous effects perhaps.

It would not appear surprising if, in future, by the means pointed out by Berthelot,* vi-

* Ann. de Chimie, III., XIV. 385.

nous alcohol = (C_4H_5O,HO) as well as wood-alcohol = (C_2H_3O,HO), and other kinds may be profitably manufactured from the so-abundant petroleum hydrocarbons.

The two methods to prepare alcohols by synthesis from hydrocarbons are based upon—

1st. In fixing oxygen upon those of the formula C_2nH_2n+2, *i. e.*, marsh gas and its homologues.

2d. In fixing the elements of water upon those of the composition C_2nH_2n, *i. e.*, olefiant gas and its homologues.

Thus Berthelot obtained wood or methylic alcohol artificially by acting upon marsh gas by chlorine, and decomposing the chloride thus obtained by means of a solution of potash.

Common or vinous alcohol may be prepared synthetically, by forming a solution of olefiant gas in oil of vitriol, which dissolves about one hundred and twenty times its bulk of the gas; then diluting the mixture, and submitting it to distillation.

Chemical Composition of Petroleum.—De La

Rue and H. Müller* have examined the Birmese naphtha or Rangoon petroleum.

It is obtained by sinking wells about sixty feet deep, in which the liquid is collected as it oozes out from the soil. At a common temperature it has the consistency of goose fat; it is lighter than water, and has usually a greenish-brown color; it has a slight, peculiar, but not unpleasant odor. It is composed almost entirely of volatile constituents, about 11 per cent. of which come off below 212° F. The fixed residue does not amount to more than 4 per cent. if it be distilled in a current of superheated steam. About 10 or 11 per cent. of the volatile matters consists of solid paraffine. When the liquid portion is agitated with oil of vitriol, some of its constituents enter into combination with the acid, but the greater part remain unaltered by this agent. In the portion which combines with the acid, benzole, toluole, xylole and cumole, have been identified, and there are several basic substances, which have not as yet been completely exa-

* Proceed. Roy. Soc., VIII. 221.

mined. The liquid, from which the hydrocarbons of the benzole series have been removed by the action of the oil of vitriol, constitutes naphtha. It may be purified by repeated agitation with sulphuric acid, washing with water, and rectification from quicklime. It is then fit for the preservation of such alkaline metals as potassium, sodium, etc.

Prof. Vohl has published an analysis of Rangoon oil. Spec. grav. 0.885.

It yielded by distillation and rectification:—

Illuminating oil, spec. grav. 0.830	40.705
Lubricating oil	40.999
Paraffine, fusing at 60° F.	6.071
Asphalt	4.605
Loss (carbolic acid ? etc.)	7.620
	100.000

The following analysis of Barbadoes tar was executed by Charles Humfrey ("Technologist," March, 1863):—

The specimen was of a dark-brown color, very viscid, with faint pleasant smell. Spec. grav. 0.940.

10 ounces gave:—

Water	½ oz.
Crude oil, No. 1, spec. grav. 0.912	5 ozs.
" No. 2, " 0.927	4 "
Coke	½ oz.
	10 ozs.

No. 1, when refined, gave four ounces of fine oil, of a pale color, and very sweet; spec. grav. 0.908. No. 2 gave 2½ ounces of fine oil, of a dark color, and some empyreumatic smell; spec. grav. 0.918.

But it is on the American continent that the most copious petroleum springs and wells have been developed within the past few years. The principal and richest oil reservoirs are situated in Pennsylvania and Canada. Other deposits have been ascertained to exist in Ohio, Western Virginia, Kentucky, New York, Michigan, and it is probable will be found in Kansas, Tennessee, Alabama, California, and Indiana.

The annexed table comprises the average composition of American and Canada petroleum according to A. N. Tate, chemist, Liverpool. These analyses were made for the purpose of ascertaining the quantities of the different products to be obtained from each. The specific gravity of the spirit and burning oil has been fixed at 0.735 and 0.820 respectively.

	1	2	3	4
	Sp. gr. 0.802	Sp. gr. 0.815	Sp. gr. 0.835	Sp. gr. 0.802
Spirits sp. gr. 0.735	14.7	15.2	12.5	4.3
Lamp oil sp. gr. 0.820	41.0	39.5	35.8	44.2
Lubricating oil . .	39.4	38.4	43.7	45.7
Paraffine	2.0	3.0	3.0	2.7
Coke	2.1	2.7	3.2	2.2
Loss	0.8	1.2	1.8	0.9
	100.0	100.0	100.0	100.0

Nos. 1 and 2. Pennsylvania petroleum of a dark greenish color, and ethereal odor.

No. 3. Canadian petroleum of a brown color and garlic odor.

No. 4. Similar to the former, and also from the United States; precise locality unknown.

Pelouze and Cahours* have made a beautiful and thorough analysis of American petroleum,

* Comptes Rendu, LVI. 505; Journ. f. pract. Chemie, Bd. 29, Heft 5 and 6. Annal. der Ch. und Pharm., Bd. LI. Heft 2.

which was exported to France where it is rectified. It is especially the more volatile constituents, boiling below 392° F., that have hitherto been examined into by these distinguished chemists. They have isolated as many as twelve distinct hydrocarbons, all homologous with marsh gas C_2H_4, or, as it may be looked upon, hydride of methyle = C_2H_3,H.

The boiling point (*i. e.*, the temperature at which it assumes a gaseous form) of the most volatile oil is a few degrees above 32° F.,* and it contains probably hydride of butyle. It is also found to be one of the products of the distillation of coal at low temperatures. The formulas, specific gravities, and boiling points of these hydrocarbons are the following:—

	Composition	Sp. gr.	Boiling Point.
Hydride of Butyle	C_8H_{10}		86° F.
" " Amyle	$C_{10}H_{12}$	0.628	
" " Caproyle	$C_{12}H_{14}$	0.669	154.4 "
" " Œnanthyle	$C_{14}H_{16}$	0.699	197.6–201.2 "
" " Capryle	$C_{16}H_{18}$	0.726	240.8–244.4 "
" " Pelargyle	$C_{18}H_{20}$	0.741	276.8–280.4 "
" " { Caprinyle / Rutyle }	$C_{20}H_{22}$	0.757	320.0–323.6 "
" " Hendekayle	$C_{22}H_{24}$	0.766	356.0–363.2

* Erdmann's Journ. für Prakt. Chemie, Bd. 89, 1863. Heft 5 and 6, p. 360.

	Composition	Sp. Gr. at 68° F.	Boiling Point.
Hydride of Lauryle	$C_{24}H_{26}$	0.776	384.8–392.0° F.
" " Cocinyle	$C_{26}H_{28}$	0.792	420.8–424.4 "
" " Myristyle	$C_{28}H_{30}$	——	456.8–464.0 "
Not named yet . .	$C_{30}H_{32}$	——	491.0–500.0 "*

In addition to the mentioned class of hydrocarbons, Pelouze and Cahours found paraffine to be a constant ingredient of American petroleum. They think it probable that there exists in it several solid hydrocarbons homologous with paraffine, forming mixtures similar to the liquid hydrocarbon series. These chemists will make this a matter of future study.

The following is the result of an analysis of Canadian petroleum made by Dr. S. Muspratt. 100 parts of Enniskillen oil yielded in distillation:—

---

* The latter four have recently been described in Comptes Rendu, LVII. 62; also Journ. für Prakt. Chemie, 92 Bd. 2 Heft, p. 99, Leipzig, 1864.

Light-colored naphtha, sp. gr. 0.794	20
Heavy yellow naphtha, sp. gr. 0.837	50
Lubricating oil rich in paraffine	22
Tar	5
Charcoal	1
Loss	2
	100

The Canadian oils and those of the State of Michigan have, like those found in South America and the West Indies, an offensive garlicky odor, which distinguishes them at once from most oils of the United States. The bad smell is chiefly owing to sulphur, and Tate traced likewise small quantities of phosphorus and arsenic in the Canadian oils.

From chemical analysis we are justified in considering the following difference between coal oils and petroleum established.

The coal oils contain the hydrides of the alcohol radicals, homologous to marsh gas $C_n H_n + 2$, in a small proportion, but the hydrocarbons of the benzole, and toluole series, in large quantity.

Petroleum contains mere traces of benzole if any at all, but is mainly made up of the hydrides of the alcohol series of hydrocarbons.

Whilst Pelouze and Cahours and others find no benzole in the American petroleum, Schorlemmer, of Manchester, on the contrary, states that it contains small quantities of benzole and toluole. Mr. Murphy, of Liverpool, could trace no benzole in petroleum, except in one or two cases in minute quantity, and then he believed it was produced by decomposition during his experiments.

Tate was unable to detect it in any specimen of the crude oil he examined, but found it in several specimens of the "turpentine substitute," also sold under the names of "benzine," and "petroleum spirits." This is obtained upon distilling American petroleum; the first product passing over has a specific gravity 0.680, and is called "kerosolene;" the next is a spirit somewhat heavier, and this has been called, though wrongly, benzine, although it may in many cases be used as a good substitute. Both the kerosolene and benzine are frequently collected together, and form the turpentine substitute above referred to. It is believed that the small amount of real benzine traceable therein results from decomposition during the process of distillation.

As Mr. Schorlemmer examined these same lighter portions of crude petroleum, we can account for his having expressed a different opinion.

We have likewise been unable to recognize benzole in some few American oils, neither could we trace it in some kinds of the naphtha or turpentine substitutes employed by house painters. A fluid sold by our druggists for removing grease spots, and marked benzine, softened but failed to dissolve India rubber, and upon examination by Hofmann's test proved to contain no real benzine. It is stated by some that in the Canadian oil benzole is found, but, strange as it may appear, we have been unable to procure any in the broker offices of New York and Philadelphia to examine it ourselves. Whether, at least, some specimens of American or Canadian oils are analogous to Birmese naphtha, in which De

La Rue and Miller found benzole, toluole, etc., we have had no means of determining thus far.

This is an important point in regard to the manufacture of aniline colors as yet generally imported from abroad. If American petroleum does not, like the Birmese naphtha, or like artificial coal oil, furnish benzole, then, of course, no nitro-benzole, the cheapest and most abundant source of aniline, would be yielded. The small amount of the ready formed alkaloid present and removed by acid washings in the refining operations, would, perhaps, not pay the trouble of isolating it.

CHAPTER VII.

REFINING OF PETROLEUM.

The general principles followed in refining petroleum are identical with those described under kerosene oil. The crude oil is distilled in a common large iron still protected by brickwork to prevent the fire from playing directly on the still. Steam pipes are inserted into the still when steam is employed in the process of distillation. With the still a coil of iron pipes or condensing worm is connected, which is placed in a vat filled with water. This is kept cold until paraffine oil begins to go over, when, to avoid its solidifying in the worm, it is kept at a temperature of 80° F.

The distillation is carried on without the use of steam until the remainder of the charge in the retort grows thick when cold. If this pitch is wished for, the operation is stopped at this point, otherwise steam is now passed into the neck or breast of the retort, which produces an outward current through the condenser, carrying over the rest of oils and leaving behind a compact coke.

Common or previously super-heated steam has also been employed, being led into the charge during distillation; this plan is of decided advantage, especially for the distillation of the heavy oils.

The still patented by Abraham Quinn, of New York City, in 1863, and which appears to be built upon an excellent plan, has the advantage of allowing the distillation to be carried on without interruption, a fresh supply of oil being constantly run into the still.

The distillate is collected in two portions. The first has a specific gravity of 0.74, and forms the turpentine substitute of our paint shops, and is falsely sold as benzine.

The second shows a specific gravity of 0.82, and is well suited for lamp oil; the balance of heavier oils is either transferred to the next charge, or kept as lubricating oil.

These two products are then each agitated for some hours with five to ten per cent. of sulphuric acid, allowed to settle, drawn off, and next agitated with water, and finally with five to ten per cent. of caustic soda liquor, specific gravity 1.40. After some hours' repose the alkali is drawn off, the oils once more washed with water, and again carefully distilled.

During all these operations the temperature of the oils ought to be maintained at about 90° F. The heavier portions of the distillate from crude petroleum form different kinds of lubricators. The best variety is that which follows after the burning oil has passed over. The residue or coke left behind in the still varies from five to ten per cent.

The garlicky odor of Canadian oils can be got rid of by the action of chemicals. By forcing them simply through such deodorizing mixtures as charcoal and sand the purpose may be attained.

Illuminating Power of Petroleum.

The following table, extracted from a lecture on artificial light by Dr. Frankland ("Chemical News," Feb. 21, 1863), shows the illuminating power of petroleum as compared with the light evolved by other substances. The table is arranged so as to show the quantity of other materials required to give out the same amount of light as would be obtained from one gallon of Young's paraffine oil:—

Young's paraffine oil . .	1.00 gallon
American petroleum, No. 1 .	1.26 "
" " " 2 .	1.30 "
Paraffine candles . . .	18.6 pounds
Sperm " . . .	22.9 "
Wax " . . .	26.4 "
Stearine " . . .	27.6 "
Composite " . . .	29.5 "
Tallow " . . .	39.0 "

The next table* gives the comparative cost of light obtained from different illuminating materials, as compared with the light of twenty sperm candles, each burning ten hours at the rate of 120 grains per hour.

	s.	d.
Wax	7	2½
Spermaceti	6	8
Tallow	2	8
Sperm oil	1	10
Coal gas	0	4½
Cannel gas	0	3
Paraffine candles	3	10
Paraffine oil	0	6
American petroleum (inferior)	0	7⅝

The table given below is taken from "Circle of Sciences," vol. i. p. 421.

Description of Oil.	Price per Gallon.		Intensity of Light by Photometer.	Amount of Light from Equal Quantity.	Cost of an Equal Quantity of Light in Decimals.
	s.	d.			
Petroleum	2	0	13.7	2.60	2.00
Sperm	7	6	2.0	0.95	20.00
Camphene	5	0	5.0	1.30	10.00
Rape or Colza	4	0	1.5	0.70	14.50
Whale	2	9	2.4	0.85	8.25

* Tate's Petroleum and its Compounds, London, 1863.

CHAPTER VIII.

HISTORY OF PETROLEUM OR ROCK OIL.

THE at present seemingly inexhaustible quantity of native bituminous oils has rendered their manufacture from any material, no matter how cheap and abundant, unprofitable. Nature distils free of charge. Indeed, it has almost ruined the whaling business; the old oil merchants sold their ships in many instances to the United States Government at the outbreak of the war, for the purpose of blockading Southern ports, and turned their establishments into refineries for the purification of petroleum, and have been obliged greatly to multiply the number of stills and vats in these. Although before the present war native petroleum was generally unknown in the country, it is not to be supposed that its discovery is new; on the contrary, it dates back to the remotest antiquity.

The liquid bitumens or petroleum, when exposed to the air, abstract oxygen therefrom, turning gradually into solid asphaltum; in this form they were in ancient times used for building purposes. In building the ancient city of Nineveh it appears that asphalt was employed as a mortar, and was probably prepared by the evaporation of petroleum. We find it also stated that the builders of Babel used "clay for bricks, and slime for mortar." (Gen. xi. 3.) It is well known that melted asphalt, together with sand, constitutes a superior mixture for roofing felt, for covering floorings, and for sidewalks. The latter process was first introduced in Neufchatel, Switzerland. Artificially prepared, tar may be similarly employed, after it has been deprived of its oil by distillation. The catastrophe of Sodom and Gomorrah may have had some connection with, if not been absolutely caused by vast natural stores of this inflammable petroleum. At least, we find immense accumulations of hardened rock oil in the centre and around the shores of the Dead Sea, where it has been converted by oxidation into rosin-like asphalt. The pieces floating upon its waters are now frequently, in the convents of Jerusalem, cut into ornaments, such as rosaries, the beads of which, when genuine, have a strongly bituminous odor. Another early and curious use of petroleum was made by the Egyptians, whose religious belief in the return of departed spirits caused them to revolt against the law of nature commanding "dust to dust," &c. Not only was every human being embalmed, but also all the animals considered as sacred. In many cases, native bituminous matter appears to have been used as a preservative, its creasote rendering it a very excellent one. In this age of steam, utilitarianism, and curiosity, many of these mummies have been borne away by marauding travellers, and, in some instances, have been used to supply the fires of locomotives. Thus has nature reclaimed her dues. It is not decided whether the Egyptians obtained native petroleum, as they might from the Island of Zante, on the west coast of Greece, whose springs are described by Herodotus, or whether they prepared artificial pyroligneous acid for the purpose of embalming.

Another celebrated locality for bitumen, which dates back beyond the historic period, is Birmah, in the Rangoon district, upon the Irrawaddy in Northern Asia. Five hundred and twenty wells sunk in beds of sandy clay and clay slate, yield annually more than 400,000 hogsheads of this oil, which is also known as Rangoon tar or Birmese naphtha. Throughout the whole empire of Birmah, and many other parts of India, it has been used for centuries for purposes of illumination, as a medicine, for rendering timber weather-proof, and for preserving it against insects. For two centuries, Amiano and other places in the north of Italy, have furnished a profusion of naphtha, and the cities of Genoa and Parma were lighted with it. Over large districts in Persia no other illuminating material is used. The phenomena it presents cause the region to be called the Field of Fire, and made Bakoo the sacred city of the Ghebers, or Fire-worshippers. On the island of Trinidad, in the West Indies, petroleum exudes not only from springs and rocks in the usual way, but it has formed a lake two to three miles in circumference; warm and liquid in the centre, where it seems always slowly boiling, but thickening as it recedes from this point, till at the margin it is cold and solid. Persons may walk upon it at pleasure when the weather is cool, but not so when it is hot. This Lake of Tar, as the inhabitants call it, is said, by travellers, to be underlaid with coal. Dr. Gesner gives the following description of it.

"The bitumen, of the consistence of thin mortar, was flowing out from the sides of a hill, and making its way outwards over more compact layers towards the sea. As the semi-solid and sulphurous mineral advances, and is exposed to the atmosphere, it becomes more solid, but ever continues to advance and encroach upon the harbor. The surface of the bitumen is occupied by small ponds of water, clear and transparent, in which there are several kinds of beautiful fishes. The sea, near the shore, sends up considerable quantities of naphtha from subterranean springs, and the water is often covered with oil, which reflects the colors of the rainbow."

In our own country, before its colonization, and perhaps before its discovery by Columbus, petroleum was known to the Seneca Indians. According to a tradition among them, its existence was revealed to one of their chiefs by the Great Spirit in a dream. He was directed to proceed to a certain spot, where he would find a liquid oozing from the ground, which should become a healing balm to his tribe. They seem to have collected it chiefly from the surface and banks of two streams, both of which have since received the names of Oil Creek; one being in Alleghany County, New York, and the other in Venango County, Pennsylvania. Along the borders of the latter there may still be seen the remains of ancient pits, which must have been dug by them to catch the exuding petroleum. They employed it for medicinal purposes, and in many religious ceremonies; but its chief use was as a medium for dissolving the rude paints with which they adorned themselves. They sold it to the early colonists as a specific for rheumatism and various other affections. The white people called it Seneca oil, after the tribe which chiefly used and bartered it, and considered it a rare and very efficacious remedy." It is recorded that the usual method of collecting it was to throw a log across one of the oil streams, and to stop the surface oil by laying blankets upon this log. When it had accumulated sufficiently, they wrung the cloths over vessels provided to hold the liquid. More than a hundred years ago, at the time of the French and Indian war, the commandant of Fort Duquesne—which stood precisely where Pittsburg now stands—wrote a letter to General Montcalm, in Canada, giving a very interesting account of a great Indian assembly in the night on the banks of Oil Creek. In the midst of their ceremonies, the oil that had collected on the water was fired, and simultaneously they shouted and danced about the flames.

Although the white settlers learned all about

the oil springs from the Indians, they gathered, perhaps, not more than twenty barrels annually, and this was solely consumed for medicinal purposes. The idea never seems to have struck them that, by deeper excavation, the supply might be much increased, nor that its quality and usefulness could be much enhanced by distillation.

Oil was first obtained by boring in 1819. In sinking wells for salt on the Little Muskingum River, in Ohio, one or two wells sunk discharged vast quantities of petroleum and gas in an explosive manner. Although Dr. Hildreth states that it was in some demand for lamps in workshops and manufactories, and that he predicted that it would be a valuable article for lighting the streets of the future cities of Ohio, yet for over thirty years it was not used in this way.

In 1845, in boring for salt water upon the Alleghany Mountains, near Pittsburg, a petroleum spring was struck, but its products were bottled and sold in drug shops at a high price.

In 1854 one of the springs on Oil Creek

was purchased on speculation, and the oil examined and reported upon, but nothing further seems to have been done until 1858, when two New Haven gentlemen resolved to continue the search for oil wells. One of them, Colonel Drake, removed to Titusville, Crawford County, and began his arrangements for boring into the rock below the bed of the creek, and in August, 1858, the oil stratum was reached, at the depth of seventy feet. A pump was introduced, which raised at first four hundred gallons, and afterward one thousand gallons daily. Business immediately assumed a new aspect in Venango County and thereabouts. The wildest fever of speculation soon ensued. Lands rose enormously in price. The length of time considered necessary for the making of a fortune was that requisite for the sinking of a shaft. The land on either side of French Creek, Oil Creek, and part of Alleghany River was perforated with wells, and the derricks for working the drills stood up in the yards and gardens of the villages as thick as masts in a harbor. The wells varied in depth from sixty to six hundred feet. The Empire Spring was of the latter depth, with a hose leading from it to a reservoir three hundred feet higher; yet the pressure of the gas which issued with the oil forced it up the whole nine hundred feet. The celebrated Phillips Well yielded three thousand barrels per day. When a well was apparently exhausted, the supply could often be renewed by drilling a little deeper.

In Ohio, not far from the Pennsylvania border, the people had noticed a strong taste of oil in the water of the vicinity, and this, after the success of the wells in Venango County, induced them to make a similar attempt. Petroleum was reached at the depth of sixty feet; and within six months after this there had been seven hundred wells sunk. Ritchie and Wirt Counties, Virginia, have also been found to produce good oil. The first attempt in New York was made about a year and a half ago, in Alleghany County, near a famous pool, which had always been known as "the oil spring." Even before the iron pipes could be driven down to the rock, the oil, mingled with water, rushed up like a fountain. The jets of gas which accompany the petroleum are often very profuse and long continued. In Chautauqua County, N. Y., they have been secured and made use of to light the town of Fredonia, and the light-house in Portland Harbor, on Lake Erie.

Previously to this time, in consequence of the usefulness of the oily products of coal introduced by Mr. Young, in Glasgow, some gentlemen in Canada—foremost among whom was Mr. Williams, of Hamilton — formed themselves into a company, and acquired the lands in Enniskillen, on which superficial deposits of a tarry bitumen occur. Their intention was to use this substance as a substitute for coal in the manufacture of such oils, it having been ascertained to contain 80 per cent. of volatile matters. It was soon discovered that, on penetrating through the bitumen into the clay beneath, the material could be obtained in large quantities in the fluid state. In 1857 Mr. Williams bored into the earth on the shore of Lake Huron, in Canada

West, and pierced a reservoir of oil. His success at once induced the sinking of wells in Pennsylvania.

The Canada oil district has surpassed all others in the immense amount it has produced, allowance being made for the fact that the number of wells is comparatively small. In 1862 the three hundred Enniskillen Wells, on Black Creek, produced, within an area of two square miles, at a minimum, four hundred gallons each per day. One of the wells spouted, for many days after it was finished, from four to five hundred thousand gallons daily. Another was only second to it in yield, and, with five or six others of remarkable richness, did continue for several months to pour forth this flood of oil, much of which was wasted for want of proper reservoirs, or an adequate supply of barrels.

Petroleum coming from different localities often differs in consistency from the fluidity of naphtha to the viscidity of tar. In color, specimens vary from extreme yellow to deep black, and some have a greenish or reddish hue.

The Canadian oils contain more paraffine in solution, and have a greater specific gravity (0.9) than American samples, which, while rich in light oils as a rule, have only the specific gravity 0.8.

The heavier hydrocarbon oils are non-explosive and safe, while the use of the lighter and more volatile ones, such as naphtha, sometimes sold under various fictitious names, as "liquid gas," "vesper gas," &c., are as dangerous as the older burning fluids, manufactured from strong alcohol and oil of turpentine, and which invariably exploded when their volatile vapors, mingled with atmospheric air, came in contact with a light. They have cost hundreds of lives in our country.

The native coal oils of Pennsylvania and Canada belong to the first class—*i. e.*, do not explode in lamps and cans. This has been determined by actual experiment.

Of the samples of American petroleum, tested by the Manchester Sanitary Commission, two formed an explosive vapor with air at 60 deg. F.; four at 100 deg. F.; three at 120 deg. F., and twenty at 150 deg. F. Nine specimens out of thirty-two were pronounced very dangerous. The British government has legislated on the subject, virtually forbidding the sale of those oils which take fire and explode at or below 100 deg. F. The proportion of these light oils in petroleum may vary from thirty to ninety per cent.

Fig. 1.

Several apparatus have been patented for the purpose of determining the temperature at which different kinds of petroleum are likely to explode.

Parrish* recommends the following construction of which we give a drawing, Fig. 1.

A thermometer F, inclosed in a capsule, passes through a cover D into the oil to be examined. A tube with wick surrounded by a chimney S and a screen E are placed opposite the thermometer. The vessel is charged first with water upon which the oil is poured next and the apparatus heated in a water-bath and the wick lighted.

Through the opening L an air current with an admixture of oil vapor is formed, which shortly takes fire with a slight explosion, putting out the flame. The temperature at which this happens is observed.

Wells often take fire by accident, and the injury consequent, both to property and hu-

---

* Proceedings of the Amer. Pharm. Association at the tenth annual meeting held in Philadelphia, 1862.

man life, is the most serious. Says one writer: "In the autumn of 1861, a well about three miles up Oil Creek was lighted by a cigar while thirty or forty people were standing about it, of whom fifteen were killed instantly by the explosion, and thirteen seriously injured. A column of fire with its head rising and falling from thirty to fifty feet continued to burn. The Little and Merrick well exploded on April 17th of the same year, just after it had been deepened, and before the boring was finished. A most terrible scene ensued; the atmosphere was filled with the sickening gas or flames, and the ground for a long distance was a sea of fire. Four wells lost everything, including 500 barrels of oil, and much other property. Only six persons lost their lives, although a large crowd stood near at the time of the explosion. All night this magnificent spectacle was continued. A steady rush of pure oil, nearly one hundred feet in height, and never ceasing its flow, burned with the noise like the roar of a heavy surf, sending volumes of black smoke up over the tops of the surrounding hills." On another occasion no less than seven flowing and three pumping wells, with thirty thousand barrels of oil, and the surrounding woods, were in flames at once. The blazing surface of Oil Creek added to the grandeur of the sight.

In some instances the gas has become ignited and burned for weeks, the mouth of the well being converted into a mighty gas burner, from which a flame has risen many feet in height. Only a few months ago, a gentleman by the name of Jacob Crowe was sinking a well on George's Creek, Fayette County, Pa., and when the drill struck the oil deposit, a powerful volume of carbo-hydrogen gases ascended to the surface, filled the atmosphere, and coming in contact with a stove in a shanty some distance from the well, a terrific explosion ensued, and flames darted into the air sixty feet high! Fortunately no one was injured, and the flames were finally subdued; but the experienced borers never permit fire anywhere near the well upon which they are working.

On Oil Creek most of the oil is found in the same stratum of sandstone; but in Canada it is often lodged in a magnesian limestone, judging from a specimen which we have examined.

John Steele, of Oil Creek valley, is said to derive an annual income of $750,000 from wells on his premises. A correspondent of the New York *Herald* states that he was ferried across the creek by an "oil prince," aged fifteen, heir to a million, coatless, hatless, and with but one suspender.

Some idea of the magnitude which the oil business has reached, may be obtained from the fact that a strip of land two miles broad and twenty in length, on both sides of Oil Creek is estimated to-day at *two hundred and fifty millions of dollars, worth four years ago but four dollars an acre.* It will be borne in mind that this is but a very small portion of the oil region of Pennsylvania alone.

The manufacture of barrels is also a good illustration of the same truth. In many parts of the State, whole communities of barrel-makers have sprung up, as in Birmah large villages of potters supply the earthen vessels there used for the same purpose. A serious loss was at first experienced from leakage. In a single journey to New York, a barrel of oil lost one-tenth, and this in spite of every precaution. The lighter oils of petroleum persistently penetrate the pores of every wood, so that in a voyage to Europe barrels often became entirely empty. Many inventions sought to remedy the difficulty, the most valuable of which is the following: a mixture of glue, glycerine, and molasses being melted, is applied to the inner surface of the barrel, and is then washed with a solution of tannin. By this process a leather-like compound is formed, which securely confines the most penetrating oils. The same compound, with the exception of the glycerine, is used in the formation of printers' rollers.

Another process of lining barrels consists in the employment of soluble glass or silicate of potash, either alone or in union with other substances.

A recent number of the "American Druggists' Circular" contains the following statements:—

Petroleum in Pennsylvania.—The Petroleum produced in the State of Pennsylvania was sold at the wells for $56,000,000 during the last twelve months, while the iron and coal of Pennsylvania only produced $50,000,000. In Philadelphia the daily sales of petroleum stocks at the regular stock exchange board are over $200,000. The number of petroleum companies organized is about one hundred and fifty, and in New York about eighty.

Petroleum in Pittsburg.—The whole number of petroleum refiners in Pittsburg is fifty-eight, with a total capacity per week of 26,000 barrels. The value of real estate, buildings, and machinery is $2,534,000, and the value of oils refined, $8,599,223, and the wages paid per annum amount to $350,000.

CHAPTER IX.

BORING OF OIL WELLS.

PETROLEUM occurs in rocks of very different geological ages, from the lower Silurian up to the Tertiary period, inclusive. In Europe and Asia these deposits are mostly confined to the more recent secondary and tertiary formations, whilst in the United States the oil wells are mostly sunk in the sandstones which form the summit of the Devonian strata. Those of Enniskillen, near Lake St. Clair, in Canada, are situated much lower in the carniferous limestone. Petroleum is seen to impregnate mostly limestones, sandstones, and shales. The rule amongst miners is, that the harder the rock may be to drill, the lighter in color, purer in quality, and smaller in quantity, is usually th oil obtained therefrom, and the softer the rock, the darker and more abundant the oil.

Wells are sunk either by persons owning the property or by companies. Some of the original owners of the land will not sell, and sink wells on their own account, generally realizing an ample fortune in a short time. In other cases wells are sunk by companies, under the direction of a superintendent, and the expenses paid from a sum set apart as a working capital. Before a well is sunk, a spot is chosen on which to commence work. This location is determined by the dip of rock, course of stream, burst of an upheaval, concentration of ravines, and other marks governing oil men, the failures and successes of others being of great benefit in making selections of spots for wells. The oil springs in Ohio, originate generally near the anticlinal lines, as seen in the accompanying diagrams, Figs. 2, 3, 4, 5, sketched by Prof. E. B. Andrews.* A derrick, resembling the frame of an old-fashioned church steeple, is erected over the spot chosen. This

* Am. Journ. Sci. and Arts, 1861. Vol. XXXII. pp. 85-93.

derrick is about forty feet high, ten feet square at the base, tapering to four or five feet at the top, where a pulley block is affixed, through which runs a rope to work the drill and hand up the boring tools, sand-pump (a tube or pump which is used to clean out the chips from the hole made by the drill), tubes, rods, etc., used in sinking or working the well. A long box, about eight inches square, is then put down till the lower end rests on the bed rock, be it one or fifty feet. This box is called a conductor, and its use is to steady the drill which works up and down inside of it. The cost of erecting a derrick is from sixty dollars to eighty-five dollars, according to its height or plainness, and the work is done by almost any man acquainted with the use of carpenter's tools. The conductor costs from $15 to $30, according to the depth at which the bed rock is reached. The drill is a heavy iron chisel with rounded and sharpened end. It is about three feet long, and weighs from seventy to one hundred pounds. It is worked up and down by means of a rod or rope

112 COAL OIL AND PETROLEUM.

Anticlinal lines, A, traced in the Oil Regions of Ohio, by Prof. E. B. ANDREWS.
Amer. Journ. of Sci. and Arts, vol. xxxii., 1861.

Fig. 2.

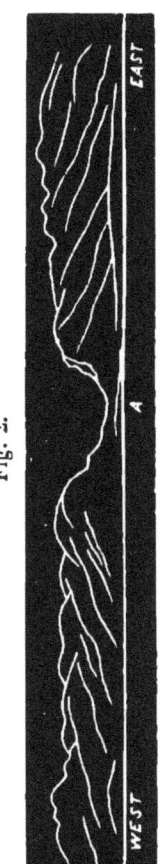

Section on the Ohio of Newell's Run.

Fig. 3.

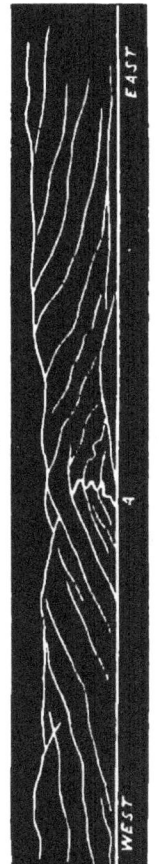

Section of Cow Creek, Va.

BORING OF OIL WELLS. 113

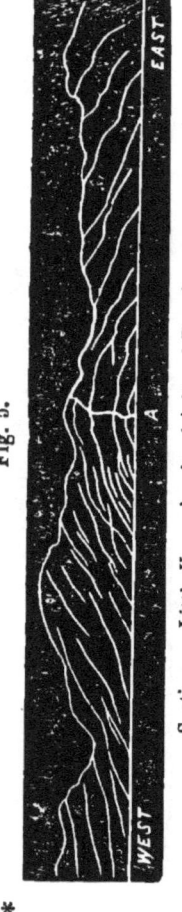

Fig. 4.

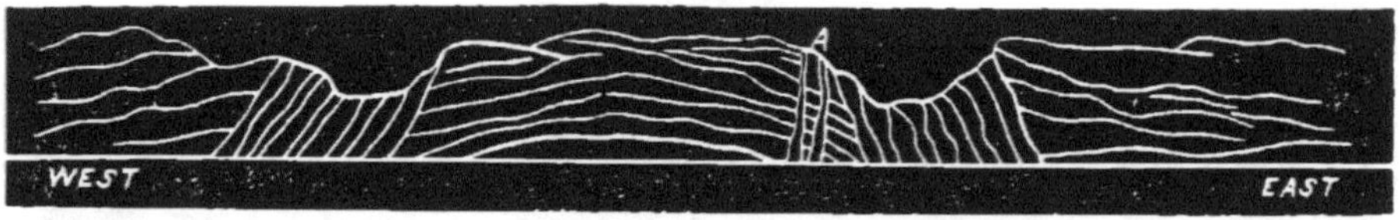

Section near Petroleum Station, North Western Virginia Railroad.

Fig. 5.

Section on Little Kanawha, in vicinity of Burning Spring Run.

Near the anticlinal lines at *A*, Figs. 2, 3, 4, 5, are gas and oil springs.

10*

attached to the upper end. This chisel is followed by a reamer, made like the drill, except the end is square. It breaks down the little irregular juttings of rock left by the drill. The reamer is followed by the sand pump, which cleans out the debris. To work the drill and other tools necessary to sink a well, spring poles, resembling an old-fashioned well sweep, are sometimes used. Spring poles are cheaper than engines at first, but not so good. The relative cost of boring is as follows: With spring poles, $3 to $4 per foot. Engine, $2 50 to $3 75 per foot. This is the price paid to men who take the contract to sink the well, the employer finding all the tools, and if the work be done by an engine, the fuel and oil to work the same. This price includes the cleaning out of the well and putting in the tubing, but not the cost thereof, which is about sixty cents a foot. The range of price is governed by the depth of the well. The usual expense is from five to six thousand dollars.

A well can be put down one hundred and

fifty feet quicker and cheaper, with springpole than with steam power. The reason is that, starting at the same time, the springpole can be erected, and the drill be down thirty or forty feet, before the engine can be set, housed, and made ready to run. But, once in operation, steam power drives ahead, passing the spring-pole drill at about one hundred and fifty feet below the surface.

For developing a new oil region, for instance, the section about Fishkill, near the Hudson River, where there are numerous unmistakable indications of petroleum, wells could be put down at less expense, and to better advantage, by spring-poles than with engines. The same rule will apply with equal force where there is a "show," but not, as in Western Virginia, that absolute certainty of great oil wealth, which but awaits the drill and pump to yield steady streams of petroleum to reward and enrich the operator.

A twelve-horse power engine costs, delivered on the ground ready for work here, about two thousand four hundred dollars; a set of tools complete, three hundred and seventy-five dollars. From the above figures men will see how much it will cost to sink a well. Tanks cost about two hundred dollars, but this expense need not be incurred till the oil is reached. The building over the engine to protect it from rain and storms costs about three hundred dollars. The barrels are furnished by the refiner, who takes the oil from the tank, pays his own cost of transportation, barrelling, etc., and keeps you supplied with empty barrels. This saves the question of transportation to parties owning or operating wells.

Rope tools are now used by all, except old fogies. They are less liable to accident, and are more convenient to draw the drill, reamer, and sand-pump, than the stiff continuous pole tools.

The primitive style of seed-bag, an old boot-leg, filled with flaxseed, which expands when wet, is still used. A better invention is demanded—one that will not provokingly give way just at the wrong time, to the delay of the works, and serious loss to the owners. These seed-bags, I remark for the benefit of those who may not know the meaning of the term, are contrivances let down the hole, outside the tube, for the purpose of keeping back the air or water, or stopping some little crack in the rock through which the drill passed. A boot-leg, filled with flaxseed, was found to answer the purpose, hence the name. Oil men will hail with delight a new invention which will be certain to do the work, and be less bungling and more easy to manage.

Oil-tools can be best and cheapest procured in Pittsburg, New York, Philadelphia, or some other place where they are to be had on order. Parties about developing oil-lands cannot be too particular in the selection of good machinery. Poor weak machinery is a nuisance, and its cheapness a curse, especially when the means for repairing are not close at hand.

We are often asked how fast a man can bore, or how many feet a well is sunk in a day. The answer depends on circumstances, according to the nature of the rock. Some days the drill

118　　　COAL OIL AND PETROLEUM.

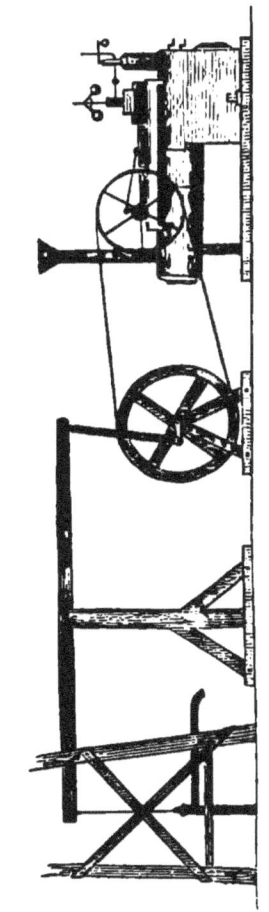

Fig. 6.

Outline of engine, derrick, and connections of one with the other.

will make fifteen or twenty feet; again, it will pound all day and not penetrate more than twenty inches. The average is about eight feet a day. The best plan is to work two gangs of men, from twelve to twelve, and without stopping drive down as fast as possible till the oil is reached. Very much of the success or failure of a company depends upon the skill, capability, genius, and business tact of the superintendent. One man works to kill time, looking more for sunset than for oil, indifferent as to all things save drawing his wages. Another man attends to business, is quick, prompt, energetic, and interested in the welfare of his employers. One good superintendent is worth twenty poor ones, and can take charge of a score or more of wells, simply requiring brains to plan and a mind to direct the labor of others. Companies cannot be too careful in this respect, as daily observation has abundantly proven.

A new process of boring is on trial at the Gillette Company's wells, on the McElhiney tract, Pennsylvania, under the management of Mr. J. T. Briggs. The process is of French in-

120 COAL OIL AND PETROLEUM.

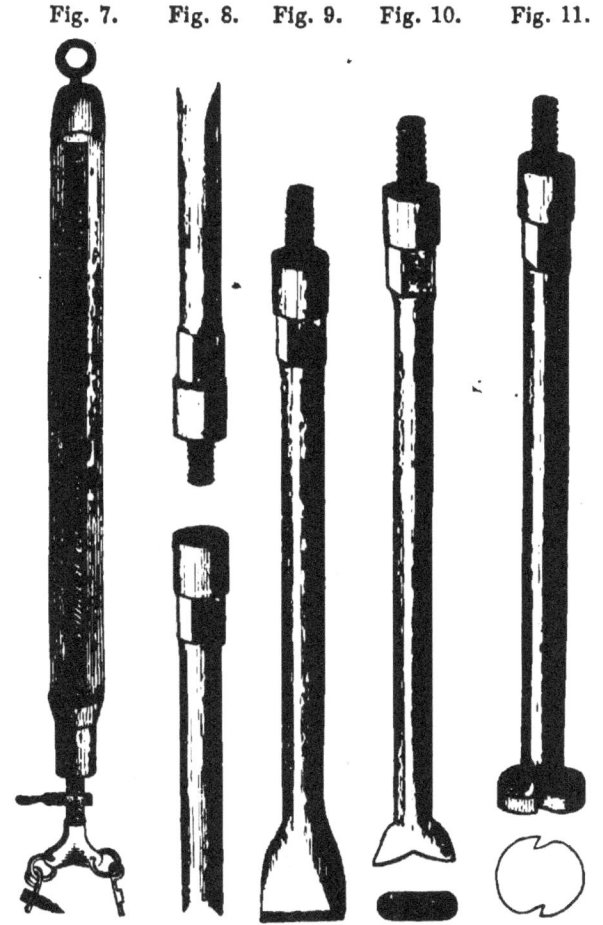

Fig. 7.   Fig. 8.   Fig. 9.   Fig. 10.   Fig. 11.

Temper-screw. Drill-stem. Drill. Reamer. Round reamer.

BORING OF OIL WELLS. 121

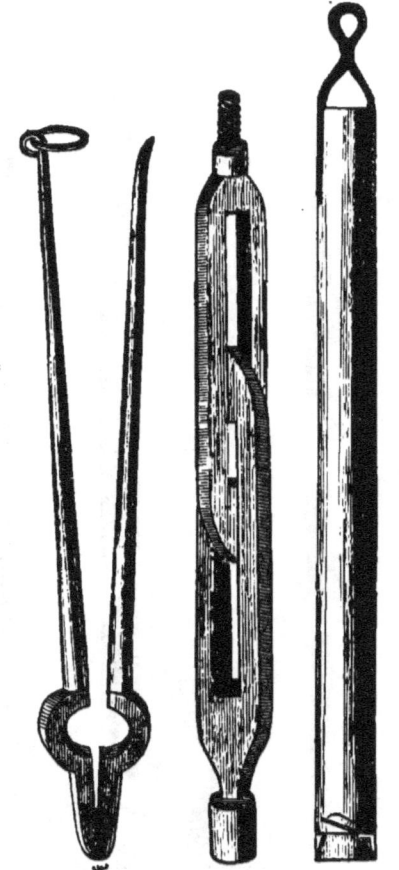

Fig. 12.   Fig. 13.   Fig. 14.

Pipe-tongs.   Jarr.   Sand-pump.

vention, and the patentee personally superintends its working. This is the first time it has ever been tested, and the progress of the experiment is watched with great interest by well owners. The principle is that of cutting out a hole instead of pounding it. The drill is circular and hollow, being a thin tube, set at its lower edge with Brazilian diamonds, of hardness sufficient to cut glass. It is connected by an iron rod to bevelled cog-wheels attached by cranks and rods to the walking beam of the engine. The surface of the upper rock being cleared, the drill sits on it and revolves with great rapidity, cutting its way down at a rate astonishing to old well borers, and leaving a central core standing. A clamp is let down which grips the core and jerks it up in the form of a perfectly smooth cylinder. Water is poured down the hole to assist the cutting process, until the natural flow from the springs cut supplies the want. The portions of the core shown exhibited the stratification of the rock, and will go far to settle some vexed questions about the strata which cannot be ascertained by the ordinary method of drilling.

Five feet of rock had been cut at the rate of four inches in five minutes, or ninety-six feet per day, when some changes were required in the machine, and it was removed for alteration. The patentee is satisfied that he can put down a well of five hundred feet deep in ten days, at no greater cost to the well owner than by the present tedious process, which takes from two to four months.

An oil well was being bored near Detroit, Michigan, and when the drill had reached a depth of seventy feet a current of gas escaped which blew out the drill and tools, weighing eight hundred pounds; blew off the shed roof of the derrick, forty-five feet high; and hurled forth a stream of water, gravel, and large stones. The workmen narrowly escaped with their lives. The water was strongly impregnated with petroleum.

CHAPTER X.

ORIGIN OF PETROLEUM.

PETROLEUM is doubtless a product of chemical decomposition, derived from organic remains, plants, and animals, whole generations of which perished and accumulated during many destructive revolutions at the various ages or epochs of our planet. As to the manner in which these oily hydrocarbons were originally produced, scientific men are still divided in opinion. Some believe that they resulted, like the artificial oils we have dwelt upon, from a dry distillation—*i. e.*, the effects upon vegetable tissue of heat, such as hot gases, or steam generated by volcanic action, untold ages before our solid earth had acquired its present thickness and stability of surface. Many other theories have gained some ground, though mostly with the vulgar, as scarcely any of them are more coherent and rational than the old oil-king's extravagant supposition of a buried shoal of whales.

That petroleum is of vegetable, or rather organic origin, is too manifest from its composition to require argument. There are, perhaps, but two opinions in regard to the manner of its production which deserve notice at our hands. The first is that the oil was derived during the first bituminization, or conversion of woody fibre into coal. The other maintains that, by a process of distillation, coal beds, or bituminous rocks, such as schists and slates, have yielded up their oily matter, which they derived from plants and animals. Even under the first hypothesis it might have been produced in two ways; for we may heat or char a vegetable substance like wood rapidly, with the total exclusion of air, and without the possibility of escape for the gaseous and liquid products, or we may suffer it to undergo a spontaneous decomposition under the same conditions of absolute confinement, and under so great a pressure that at the commencement of the process the products evolved might be retained. In both cases the same result would be attained, *i. e.*, the formation of stone coal, oils, and gases, although in the latter case sufficient time—perhaps thousands of years—must be granted. Experiments of the former kind, *i. e.*, charring of wood in hermetically sealed heavy cast-iron vessels, were actually executed; not so much to explain the origin of coal oils, then (1841) scarcely noticed, as to establish the experimental proof of the formation of stone coal from vegetable matter. Contrary to the usual supposition of a microscopic cellular structure, cannel and similar varieties of coal show, by their conchoidal fracture and total want of an organized or cellular constitution, that they must have been in a softened or even liquid condition. Dr. A. Petzholdt* succeeded in preparing three varieties of coal in this way, viz: coke, charcoal, and cannel coal. The

---

* Dr. Alexander Petzholdt. Ueber Calamiten und Steinkohlenbildung. Dresden and Leipzig, 1841, pp. 17–28.

latter result was obtained when the products of distillation were entirely retained, or could only have escaped through the pores of the cast-iron boxes. In this case the space occupied by the coal was one-half that of the wood; the resulting product was dense, burned readily in a candle flame, and was destitute of vegetable structure; from which he concludes that a decomposition of vegetable matter is possible in which the carbon is dissolved in the gaseous and liquid products formed at the same time. Monsieur Barouler, by placing vegetable material in an apparatus made of wet clay, and capable of being strongly compressed, and exposed for a long-continued time to a temperature ranging from 392 deg. to 572 deg. F., produced ordinary coal.

Against this theory of oil formation by the original bituminization of vegetable matter, it has been objected that the presence of bituminous coal beds, made of land or fresh water plants, should be accompanied by corresponding quantities of oil. Such is not found to be the case, since the oil, instead of being found in contact with coal deposits, occupies usually cavities of overlying rocks, perhaps produced long after the coal; and further, it is often met with miles away from any coal field. It has also been held that upon such a supposition it would be difficult to account for the large volumes of inflammable gases which exist with the oil.

In answer to these objections, it may be urged that the immense pressure to which all these substances were subjected would naturally operate to compel the condensed volatile products to seek as high a level as they could reach, and the consequence would be, that, having penetrated the shaly formations, already, perhaps, charged with the heavier hydrocarbons, they would at last find a resting place as petroleum, and remain pent up until the drill of the oil speculator gave them vent. By the same cause the oil might have been forced miles away from the place of its original distillation; although it is safe to affirm that in most localities of the United States where petroleum is found, the deposit has some geographical connection, at least with coal-bed regions, whether these belong to the oldest coal formations, as those of the Devonian and carboniferous systems, or to the much more recent oolitic and tertiary age. It is said that tertiary coal beds underlie the Rangoon oil wells. Tertiary lignites abound in Trinidad, Lombardy, and Middle Asia. As an exemplification of the pressure to which this putrescent flora was submitted, we may instance the fact that in the most recent deposits of lignites, stems of trees, upon which one hundred annular rings could yet be counted, were so flattened that one diameter exceeded the other four to eight times.

It is well known that, in the common peat bogs of the present day, the cryptogamous plants composing them give off at their first stages of decay considerable quantities of a combustible hydrocarbon known as marsh gas, together with nitrogen, some carbonic acid, and water.* Now the ancient coal measures originated from a terrestrial flora, in stagnant waters,

* Beiträge zur Erkenntniss der Kusammensetzung und Bildung des Forfes, von Dr. Justus Websky, in Journ. f. Prakt. Chemie, Bd. 92, 1864. Heft 2, p. 65.

where vast bogs of gigantic cryptogamous plants—as tree-like ferns, club mosses, horsetails (calamites), &c.—were buried, and afterwards converted by ages of time, tremendous pressure, and the agency of heat (at least in the case of anthracite), into coal. It is not possible to conceive that these processes could have gone on without the liberation of gaseous as well as liquid products in large quantities. Chemists have separated from petroleum as many as twelve volatile fluid hydrocarbons, homologous with marsh gas (C_2H_4), *i. e.*, forming, as it were, a progressive series of that first and lowest member. The rate of increase being C_2H_2, as C_2H_4, C_4H_6, C_6H_8, &c., &c. An admixture of olefiant gas (C_4H_4) as it is usually met with in gas springs, could be easily accounted for, since we know that an elevated temperature acting upon vegetable matter will produce it. Many of the heavier hydrocarbon oils entering into the composition of petroleum are homologues of the hydrocarbon C_4H_4, the last and highest member of which is probably paraffine, $C_{4+n}H_{4+n}$.

It might be objected to the formation of these gases that there was no room for them to expand under such a pressure, and that consequently they could not originate. But we know from recent experiments of Deville, Troost, and Cailletet, that combustible gases at a high temperature will even penetrate heavy iron tubes, such as gun barrels; how much less resistance would certain rocks offer?

An important item in accounting for the source of heat is the established fact that minerals, salts, &c., in the act of crystallization set free a great deal of latent heat. The amount is at times so great that lava nearly cold has been seen to become again glowing. It is also interesting to know that mineral masses may dissolve large quantities of gases before becoming solid; thus Cailletet and Pilla observed that cold lava, after the eruption of Vesuvius in 1861, evolved marsh gas and hydrogen. From this it likewise follows that the atmosphere of the crater inclosing the melted lava consisted in a considerable degree of these gases, and that they were taken up in the same manner as oxygen is absorbed by melted silver, again to be set free as soon as crystallization begins.

The annexed table* shows how, in the process of bituminization, the proportion of oxygen decreases as we proceed towards older carboniferous formations. It has been prepared from a large number of analyses, and will give at a glance the comparative composition of all these varieties of carbonaceous deposits, from wood down to anthracite coal. The carbon is represented by 100 in all cases, so as to enable us to compare the successive changes in composition which take place from wood to coal. The small amount of nitrogen and of ashes are left out in the statement:—

* From a Course of Lectures by Dr. Percy, at the London School of Mines.

	Carb.	Hydrog.	Oxygen.
Woody tissue	100	12.18	83.07
Peat	100	9.85	55.67
Lignite	100	8.3	42.42
South Staffordshire coal	100	6.12	21.23
Steam coal from the Tyne	100	5.91	18.32
Semi-anthracite coal from South Wales	100	4.75	5.28
Anthracite from Pa.	100	2.84	1.74

Finally the chemical qualities of petroleum prove that it must have been produced by a process analogous in result to the dry distillation of peat. The native oils of the United States differ considerably from the artificially prepared from bituminous coal, for they yield with nitric acid little or no artificial oil of bitter almonds (nitro-benzole), or the precious aniline dyes mentioned in connection with coal oil. They are composed mainly of volatile hydrocarbon oils obtainable at low temperatures from turf. Pelouze and Cahours remark that the total absence of benzine or any of its homologues in American petroleum, would seem to indicate

12

that this oil could not be derived from coal unless this latter had undergone a decomposition very different to that which takes place when it is submitted to distillation.*

The other theory which regards petroleum as derived from bituminous shale or coal by a process of distillation has also its objections; the chief of which is, that such coals in our oil regions furnish no evidence of having lost any of their normal quantity of bitumen. At Petroleum, Ritchie County, Virginia, where strata have been brought up by an uplift from several .hundred feet below, seams of cannel and bituminous coal appear, which, when analyzed and compared with Nova Scotian or English coals, have lost no bitumen—a fact all the more surprising when it is remembered that freshly-mined coal undergoes even at a temperature little above that of the atmosphere, but under increased pressure, the first step of bituminization, *i. e.*, disengages marsh gas or the

---

* This sweeping conclusion seems inadmissible, considering Schorlemmer's analysis of cannel coal oils.—*Journ. Chem. Society*, xv. p. 419.

so-called fire-damp so dangerous to miners. As the temperature increases, liquid hydrocarbons begin to appear.

We have indeed good reasons for believing that the bitumen associated with schists and shales is rather the result than the cause of petroleum, *i. e.*, that bitumen consists of hardened drops of the latter.

Prof. H. D. Rogers' observations are important in this connection.

He states that the amount of volatile substances in the Appalachian coal fields decreases in passing from west to east; and that, at the western limit, where the strata are still horizontal, the proportion of volatile matter may reach forty to fifty per cent. On the eastern side, where the strata have been actually turned over, the coal contains only from six to twelve per cent. Sir Charles Lyell, commenting on these observations, remarks: " There is an intimate connection between the extent to which the coal has parted with its gaseous contents and the amount of disturbance which the strata have undergone. The coincidence of these phenomena may be attributed partly to the greater facility afforded for the escape of volatile matter, when the fracturing of the rocks had produced an infinite number of cracks and crevices, and also to the heat of the gases and waters penetrating these cracks, when the great movements took place which rent and folded the Appalachian strata." According to the theory under consideration, we should expect to find oil in immense quantities wherever coal measures have parted with their bitumen; whereas, exactly the reverse is proved to be the case—it being borne in mind that deposits of oil are found only in the western portion of the coal fields, and that, in the eastern, where the coal is almost entirely anthracite, none have been discovered.

There being, probably, no organic deposit, either entirely animal or vegetable in its nature, but all more likely being composed of both, it is safe to conclude that bituminous oils are of a mixed origin in this respect. In Canada, New York, and perhaps in Kentucky, where oil is found in the Devonian rocks below the old red sandstone, it has been suspected to be mostly of animal origin, because these strata were formed long before the oldest coal measure, and exhibit no remains of a land flora. In cavities formed in the rock by some fossil animals, as the huge chambered shelled orthoceratites, some of which were many feet in length, considerable quantities of petroleum have been found, but so fetid as to be offensive. Most rocks have been formed by marine depositions of earthy matter, inclosing, in great profusion, the remains of those extinct animals which peopled the ancient oceans. These fossil shells are distributed everywhere, from the dawn of paleozoic life up through each succeeding age. The same is relatively true in regard to plants, even if that marine and scanty flora should not have been preserved, or failed to leave traces of its existence behind. It is a law of nature that vegetable life precedes that of animals; or, differently expressed, all animals are slavishly bound upon the existence of plants, since all derive their food directly or indirectly from them.

12*

The reign of plants in the carboniferous era commenced when land and water no longer struggled for predominance, for they are essentially terrestrial or fresh-water formations, presenting the appearance of huge swamps, composed, with few exceptions, of plants which might flourish either in or out of stagnant water.* The period must have been one of long, uninterrupted, and quiet growth; the climate a warm and uniformly tropical one, and the atmosphere probably highly charged with water and carbonic acid—conditions very favorable for the rich development of plants, though unsuitable for the respiration of higher animals. Indeed, at the present time, in the damp and warm climate of the South Sea Islands, ferns and equisetaceous plants assume a tree-like habitus. The decay of this very extensive cryptogamic flora, extending far up to the north, must have been slow, and have taken place generally under water. The oxy-

---

* Lesquereux, Am. Jour. Sci., vol. 32, 1861; and Oswald Heer's Urwelt der Schweiz, Zurich, 1864.

gen being thereby excluded, the carbon would be preserved.

In Southern Ohio and Western Virginia the petroleum is apparently found in the coal measures themselves; but the wells have often to be sunk through them into the sandstone and slates below before they become productive.

From the black shales, which immediately overlie the corniferous or Devonian limestone, the oil springs of Canada West issue; and from this fact the origin of the petroleum of these regions is held by the best geologists of the Province to be principally animal.

Prof. J. P. Lesley writes:* "The connection of the oil regions with the coal basins of Western Pennsylvania and Virginia, Eastern Ohio and Kentucky, is, in good measure, a geographical deception. The Oil Creek rocks, dipping southward, pass 500 or 600 feet below the coal measures. The nearest coal bed to the more northern springs occurs on the highest hilltops, many miles away. The hills in the vicinity of some of the wells are capped by the

* Article on coal oil in *Agricultural Report*, 1862, p. 443.

conglomerate base of the coal measures at least a hundred feet thick. The shales and sandstones of the valley belong to formations X, IX, and VIII, descending, called by the New York geologist the Catskill, Chemung, and Portage groups, extending over all the southern counties of Western New York. The southern dip carries down these oil-bearing rocks, and the wells must deepen in the same direction. Mr. Ridgeway reports (July 10, 1862) the lowest oil-bearing sand rock, capping the hills near Waterford, on Le Bœuff Creek, and the same sandstones appear on Big French Creek, full of plant remains.

"The following wells show the dip in a well-marked manner: The Phillipps Well, on Oil Creek, is 460 feet; the Brawley Well, at the mouth of Cherry Run, 503 feet; the Cornwall Well, 530 feet; the Avery Well, over 700 feet; and at Titusville he estimates the proper depth at 1,000 or 1,200 feet.

"In the Mahoning Coal Oil region in Western Pennsylvania and Eastern Ohio, near the line, the three oil-bearing sand-rock strata are beneath the lowest coal bed."

Sir William Logan* has pointed out that a line drawn through London, Burlington Bay, Zone, and Chatham, marks the summit of a flat, anticlinal arch (resembling a house roof), upon which the principal oil fields are situated. The same strata in which it is found dip away until in Michigan, on one side, they are proved to be one thousand feet below the surface, and in Pennsylvania, on the other, they underlie the great coal measures. It will thus be seen that the surface rocks of the oil region in Canada are the same upon which the great layers of the true carboniferous or plant-producing era are based in other localities. In Canadian rocks of the Silurian and Devonian ages, bituminous beds, and evolutions of gaseous and liquid hydrocarbons occur throughout the whole system; and in the Hudson River group of rocks, in which there are but slight traces of vegetable life, these oils have been obtained. The upper beds of carboniferous limestone, and the entire mass of Hamilton shales are charged to excess with organic (and

* Canadian Journ. New series. Vol. vi., 1861, p. 319.

mostly animal) remains. It is believed that at the time when the region in question was covered to a great extent by the waters of the ocean, a few species of aquatic plants, with various animals of a low order, such as crinoids, which grew on a stem like a vegetable; brachiopods, which were a kind of shell-fish; and trilobites, a large crustacean, and many others, flourished in wonderful profusion. As the floor of the sea gradually sunk with the cooling of the earth-crust, successive generations of these must have been buried in the waters, and covered by mineral and earthy deposits. Under such circumstances a slow, dry distillation might gradually take place, the products of which are preserved for the use of man in the innumerable fissures of the rocks. The organic matter, scattered in such abundance along the shores, would commence to decompose in the ordinary manner under the influence of air and moisture: but when, after putrefaction, it was covered with layers of sand or calcareous mud, and thus removed from atmospheric action, the resulting gases would be confined as in a closed retort, and the carbon and hydrogen, being greatly in excess of the oxygen, would enter into such combinations as we find subsisting in petroleum and the various hydrocarbon gases. That the Canadian oils furnish no conclusive proof in their component parts of an animal origin, is not to be urged against the evidence afforded by the strata in which they are found. Animal and vegetable tissues, when confined without the presence of oxygen, will give products quite closely resembling each other. Indeed, chemists have long since proved that some of the lowest classes of mollusks now living, to which the name of Tunicata has been given, have much the same composition in their mantle or covering as woody fibre. It is well known that the smell of Canadian oils is far more offensive than those of the United States. The same is the case with Michigan oils, and may apply to Kentucky and other oils found in sub-carboniferous strata. The subjoined extract from a recent newspaper contains many facts in support of the statements already made.

"An active general interest has been awakened with regard to the petroleum region of New York, and it is safe to say that the time is not far distant when the Pennsylvania oil wells will be paralleled, if they are not distanced, by those actively operating in the former great State. In Cattaraugus County several wells are now working, and in Ontario County boring for oil is now being extensively carried on. In order to furnish our readers with some information in regard to the geographical characteristics of the Genesee valley oil region, we give below two engravings showing the position and comparative thickness of the various strata at a point in this region where two wells are now going down. We are permitted, through the kindness of Walter S. Hicks, Esq., of Bristol, Ontario County, who is engaged in sinking several wells, to use in our description portions of a private letter addressed to that gentleman by Professor James Hall, the State geologist. Some thirty wells are now going down in the locality referred to, and great excitement is said to exist throughout the whole region.

ORIGIN OF PETROLEUM. 145

Fig 15.

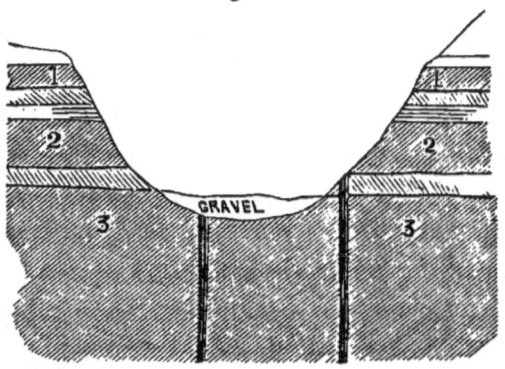

SECTION SHOWING THE STRATA AND FORMATION OF THE HILLS BOUNDING THE VALLEY.

1. Portage group.   2. Genesee slate.   3. Hamilton group.

"Professor Hall says: 'The Hamilton group is, for the most part, a close, compact shale, with few bands of calcareous matter, and few open fissures.' Therefore, he thinks, oil is not to be expected in any paying quantity until the Marcellus shale, in which there is often a thick band of limestone, is struck; but he concludes, it will be necessary to go down to the corniferous limestone, or even through it, to the Onondaga salt group below, before finding a rock sufficiently fissured or cavernous to con-

13

COAL OIL AND PETROLEUM.

Fig. 16.

Horizon of a well on Wilder gulley.

Horizon of a well on Hick's farm.

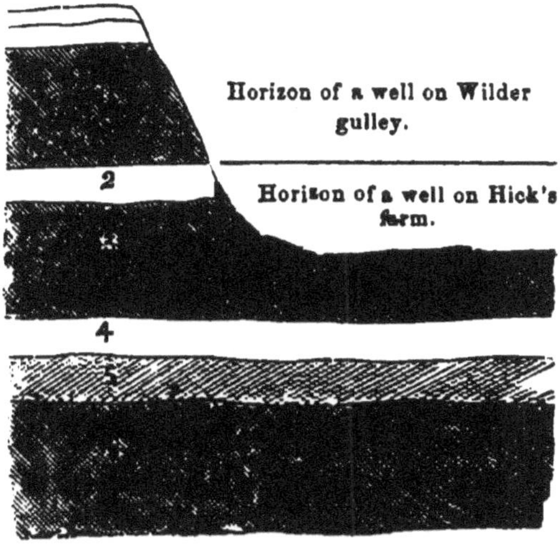

SECTION SHOWING THE STRATA AND FORMATION FROM THE SURFACE TO THE POINT WHERE OIL IS LIKELY TO BE FOUND IN PAYING QUANTITIES.

1. Portage group, forming the hills—thickness in all, from 600 to 800 feet.

2. Genesee slate—150 feet.

3. Hamilton group, consisting of calcareous limestone, with calcareous bands and a black, bituminous shale at base.

4. Marcellus shale—thickness, together with that of the Hamilton group, 600 to 800 feet.

5. Corniferous limestone, 50 to 150 feet.

6. Onondaga salt group, consisting of impure and unequally bedded limestones, beds of plaster, &c., entire thickness, 800 to 1,000 feet.

tain any considerable reservoirs of oil. Professor Hall is also of the opinion that, at the same time, more or less salt water, and perhaps strong brines, will be found. He adds that there are sometimes small masses of sandstone just below the corniferous limestone, and if one of these should be struck, a reservoir of oil would be penetrated. Large quantities of water are to be expected before oil is reached."

Having thus spent considerable time in the notice of such methods of accounting for the existence of petroleum as have gained most ground, let us rapidly observe the manner in which it is stored away beneath the surface of the earth, and also some of the best "oil signs."

The opinions of geologists, briefly stated, are as follows: That the bituminous vapors originating from organic deposits have penetrated fissures and cavities of overlying rocks caused by erosion and uplifting. As bubble after bubble passed through the water generally collected in these fissures, a portion, being condensed into oil, would float upon the surface of the water, and the remainder, in the form of inflammable gases, would occupy the top of the fissure. Where the gas finds an outlet, there is produced a "gas spring;" where the water escapes it carries the oil with it, and an "oil spring" results.

That the oil is accumulated in fissures in the rocks, and that these fissures are more or less vertical, *i. e.*, narrow and upright cavities, is decisively proved by the following facts:—

1. The oil in the same immediate neighborhood is found at very different depths, and it is very seldom that two adjoining wells strike it at the same distance beneath the surface.

2. The oils of wells very near each other may show a great difference in density, color, &c.

3. The oils from two wells not many rods apart may not only vary in specific gravity, but the deepest well may contain fresh water, while the other casts up salt water mingled with the oil.

From these observations it is evident that the oil is not found in horizontal lakes or re-.

servoirs, but in separate, distinct, and more or less vertical cavities. Several of these may be connected, however, by some channel, and thus the supply be quickly replenished when one spring or well has ceased to flow.

According to the point of the fissure struck in boring, different material may be yielded. If it is pierced near the top, gas escapes with violence, but subsequently, as the water tends to rise higher in the space thus vacated, the oil is also carried to the end of the boring, and may be pumped out. If, however, the water should enter more rapidly than it is removed, the oil may be floated to the higher parts of the cavity, and cannot be recovered until the latter is pumped away. If the middle portion of a fissure be tapped, oil appears at once in the well, and may even be forced up violently by the accumulated gas pent up above.

In locating oil wells, the following practical hints may prove useful. There exists no such thing as a specific *oil rock*, or stratum indicating the presence of oil; as certain distinct

150   COAL OIL AND PETROLEUM.

geological formations indicate and are found associated with some of the precious minerals

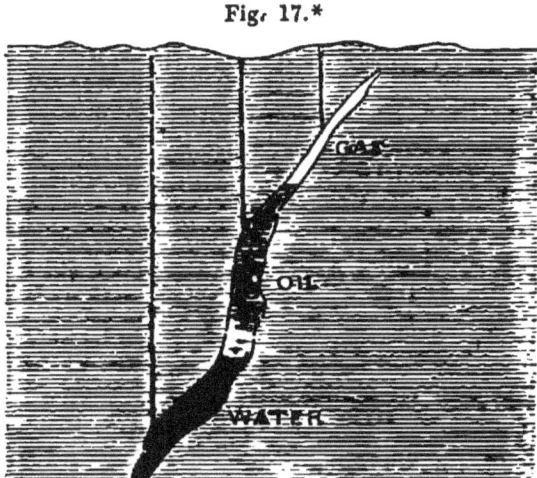

Fig. 17.*

and metals. Oil may be struck in *all kinds* of rock, for the fissures containing it may extend vertically through many different strata. Nor can anything be determined before hand from the shape and proximity of hills, as is the case in boring artesian wells.

* Rock Oil, its Geological Relations and Distribution, by Prof. Andrews. Amer. Jour. of Science and Arts, vol. 32, 1861, pp. 85–93.

The flow of oil would seem to be caused exclusively by the pressure of pent-up gas, and not, as is the case with artesian wells, produced by the weight of water, the head of which is higher than the issue. If the position be true, that the oil is generated from coal and other bituminous strata, we should expect to find it at a greater or less height above the latter; and indeed Professor Andrews has found oil springs high up on hill sides, one hundred feet above the valley below.

The presence of oil on the surface of water is no sure sign of its existence beneath, because it may have been transported from a distant source. A more favorable indication is the accumulation of viscous, hardened oil, as seen in many localities, and called "tar springs." These, as also gas springs,* are valuable omens.

It should be remembered that whatever may be the common idea, no oil well is inex-

---

* In locating wells it should be remembered that oil and gas springs may have risen to the surface in a tortuous and not in a vertical direction.

haustible, for no fissure, or system of fissures, can endure the drainage of a steam-pump for any great length of time.

The experience of other regions has revealed the fact that some ancient oil springs have ceased to flow, whilst others, like those of Birmah and Persia, continue to flow as they have for ages past apparently.

We find in the *N. Y. Journal of Commerce* of April 27, 1865, the following just remarks on this subject of the uncertainty of oil wells:—

"There is nothing more uncertain than an oil well. If it graciously chooses to do so, it yields oil in great or small quantities; but if its will be perverse, no amount of coaxing will draw out the oleaginous treasure. Neither can it be safely predicted that a well which yields a dozen barrels to-day will yield one to-morrow. They have a well in Athens County, Ohio, says a correspondent of the *Cincinnati Gazette*, which, when sunk to the oil-rock, suddenly spouted forth such a stream of oil as to threaten the oleaginous overthrow of all that county. But after a while the flow subsided, and from the reservoirs hastily constructed in the ground one hundred barrels of oil were afterwards collected; since that there has been but one well that 'blows,' as the workman express it. That well, soon after being opened, began to throw out oil to the height of twenty feet, but only for a short time. Last summer, for a considerable time, it observed a regular period of twenty-four and one-half hours between these 'blows,' each day the phenomenon occurring half an hour later than on the previous day. At last when the time for its 'blowing' reached far into the night, it lost its regularity, and now seems to be governed by no law, but still keeps 'blowing' almost daily. When supplied with a pump and engine, it stubbornly refused to yield at all, and the engine and pump were taken away, when the well resumed its 'blowing.' From another well nothing was got for several weeks but water; after being sunk and giving the usual indications of oil, it was tubed and prepared for pumping. To the dismay of the company, only water was obtained as the result of the first day's work. Water flowed abundantly the second day, but no oil. The third day was but a repetition of the first two, and the well was about to be abandoned. One member of the company, however, suggested the idea that the water might be exhausted by constant pumping, and that then oil would be obtained. Being an obstinate man, his counsel prevailed, and day and night, without ceasing, the tireless engine pumped water a whole fortnight. Still no oil. Another fortnight and only water appeared; another and another, when lo! the flow of water ceased, and the flow of oil began. Eight weeks of constant clinging to a theory brought a triumph to the obstinate member, and a reward to the whole company. On one occasion since, when for some reason the engine was stopped for half an hour, it required nearly twenty-four hours pumping to clear off the water. Again when a belt broke and caused a stoppage of fifteen minutes, the same thing occurred. The well is now kept constantly running, and produces from twelve to fifteen barrels a day."

GENERAL VIEW OF THE GEOLOGICAL DISTRIBUTION OF PETROLEUM IN THE UNITED STATES AND CANADA.*

The lowest geological horizon or rock stratum in which petroleum is found in large quantity, is in West Canada at Enniskillen. The oil is in the corniferous limestone. This formation in the United States is, in its maximum, about 350 or 400 feet thick.

Immediately overlying the limestone is the marcellus shale, which is also highly charged with bitumen. It is about 50 feet thick in Canada. These two rock-formations, then, which in Canada are not over 150 feet in thickness, are the reservoirs holding rock-oil in that country. Ascending in the geological scale and passing over into New York, the next stratum of rock yielding bitumen, oil, and gas, is

---

* From an article in the Scientific American. Compare, also, History of Petroleum, by T. S. Hunt, in the London Chem. News, July, 1862; also Natur. History of New York, Part IV. Geology, by James Hall, Albany, 1843.

there known as the Hamilton Group, about 1000 feet thick. The oil springs of Western New York, along the banks of its numerous lakes, are mainly in this group of rocks. They have as yet yielded oil only in small quantities for medicinal purposes, but promise very fair.

Above this group succeed black shales, known as the Genesee Slate, 300 feet thick. The wells of Mecca, Ohio, and others of that region are most probably in this rock. Above the Genesee Slate comes in the Portage Group of slates and sandstones, 1700 feet thick. The deeper wells of Oil Creek, Pa., will reach the sandstones of this group. Still above lie the rocks of the Chemung Group, which are mainly composed of thin bedded slates and limestones. In its maximum it is 3200 feet thick, but in Western New York and Pennsylvania it is much thinner, being only about 1000 feet thick. Much of the oil of Oil Creek is from this group; 400 or 500 feet of it are seen in the cliffs and hills of Oil Creek, the Alleghany River and its tributaries above, and in Venango County.

Measured in the maximum development of all the rocks enumerated, we find between the oil of Canada and Venango County, Pa., 6000 to 7000 feet of sedimentary rock, all of which bear the appearance of having been deposited in sea-water. The entire group of rocks enumerated are known as the Devonian Series in England. The oil springs of Eastern Canada and New Brunswick, along the Gulf of Newfoundland, are in the upper members of this series.

Leaving for the present those portions of the United States where oil has been most successfully found, and before coming into the geological strata of the thick and heavy oils, we have on the eastern flanks of the Appalachian Mountains in Pennsylvania and Virginia, 5000 feet of the Catskill group of rocks. (Ponent of Prof. Rogers.) Lapping around the southern outcrop of the coal measures of Tennessee, Kentucky, and Illinois, there are 200 feet of the lower carboniferous and 300 feet of the middle carboniferous. (Umbral of Rogers.) A total in the aggregate, as measured in Nova Scotia and the United States, of 1500 feet.

Throughout the whole of the series oil and gas springs are found.

We now come into the true coal measures. These are divided into lower, middle, barren, and upper measures, a total of the bituminous portion of 2500 feet.

The lowest member of the coal series caps the highest hills, near the mouth of Oil Creek, and lies about 600 feet above the bed of the creek, or 1300 feet above the third sand-rock, which is the most abundant oil-producing stratum.

At the Kiskiminetas, Slippery Rock, Butler Co., Pa., and Smith's Ferry, oil is in the lower coal measures—800 feet thick. High up Kiskiminetas and on the Monongahela River, oil is found in the middle coal series 1000 feet thick. At Marietta, Ohio, and in the oil region around the strata of the upper coal, are the productive series.

To conclude, then, oil is found through 24,000 feet of rocks, as measured vertically in the geological scale, and geographically from Nova Scotia to Lake St. Clair, and from Virginia to Tennessee River. The geographical area covered by the oil-bearing group of rocks in the United States, Canada, New Brunswick, and Nova Scotia, cannot be less than 200,000 square miles.

CHAPTER XI.

PREPARATION OF ANILINE DIRECTLY FROM COAL TAR; AND ITS PROBABLE ORIGIN.—ARTIFICIAL PREPARATION OF ANILINE FROM BENZOLE; TRANSFORMATION OF THE LATTER INTO ANILINE.—PROPERTIES OF ANILINE.—CHEMICAL TEST FOR BENZOLE.—COLORING PRINCIPLES DERIVED FROM ANILINE—THEIR MODE OF PREPARATION AND APPLICATION IN DYEING.

AMONGST the most interesting discoveries of modern times, must be ranked that of preparing from an oily substance named aniline, a whole series of the most superb dyes, causing almost a revolution in the art of dyeing and printing. Aniline or phenylamine $= C_{12}H_7N$, or $= N \begin{cases} H_2 \\ C_{12}H_5 \end{cases}$ constitutes an artificial organic alkaloid analogous to ammonia $= NH_3$; the radical phenyle = $C_{12}H_5$ replacing one equivalent of hydrogen. It may be obtained in manifold ways for laboratory purposes thus:—

Powdered indigo digested with a strong solution of caustic potash and subjected to distillation yields a small amount of aniline. From the indigo plant (*Indigofera anil*) it derived its name, Aniline, synonymous with Kyanole, Benzidam, and Crystalline.

It is now always prepared for industrial purposes from coal tar.

1st. It may be procured directly from coal tar, *i. e.*, as a biproduct in the manufacture of the heavier photogenic oils and of paraffine. These oils are, as stated elsewhere, decolorized and purified by means of acids and alkalies; now the ready formed aniline forms, together with leucoline, picoline, &c., a portion of the dark brown acid wash.

Preparation of Aniline directly from Coal Tar.

The ready formed aniline is extracted from coal tar oils as follows, according to Hofmann:—

The mixture is agitated with rather concentrated hydrochloric acid, and the supernatant layer of oil separated from the acid liquor; this latter is brought in contact with new quantities of oil until nearly saturated, so that only a slight acid reaction prevails.

This solution is now placed in a copper still and treated with an excess of milk of lime, and distilled.

The condensed products in the receiver are collected, carefully dissolved in hydrochloric acid, and filtered through coarse filtering paper, which retains that portion of indifferent oil floating on the rest of the liquid.

The filtered solution is concentrated in a water-bath and treated with hydrate of potash or soda, whence the basic oil or aniline separates and swims on the top of the alkaline liquor. It is removed with a pipette, brought in contact with melted hydrate of potash to deprive it of water, and rectified. The receiver is changed and the operation suspended as soon as the product passing over furnishes no longer, upon addition of a few drops of chloride of lime solution, the purple reaction of aniline.

Origin of Aniline in Coal Tar.

The answer to the question as to how this alkaloid originates in coal tar seems to be this. It must be derived from carbolic acid or the hydrated oxide of phenyle found therein to the amount of eight to ten per cent. For if this body together with ammonia gas is passed through a red-hot tube, the following reaction takes place:—

$$\underbrace{C_{12}H_{5}O + HO}_{\text{Hydrated oxide of phenyle or carbolic acid.}} + \underbrace{NH_{3}}_{\text{+ Ammonia.}} = \underbrace{\left\{ \begin{array}{c} C_{12}H_{5} \\ NH_{2} \end{array} \right\}}_{\text{= Aniline.}} + \underbrace{2HO}_{\text{+ Water.}}$$

It has been shown that a small amount of aniline may be produced by saturating carbolic acid with ammonia, heating the mixture in a hermetically sealed glass tube to 572° F. by means of an oil bath. Indeed, if we heat in a test-tube a little carbolic acid previously saturated with ammonia, we obtain upon the addition of a few drops of chloride of lime solution the blue reaction of (impure) aniline.

On the contrary, aniline may, when acted upon by nitrous acid $=NO_3$, be decomposed into carbolic acid.

$$C_{12}H_7N + NO_3 = C_{12}H_6O + HO + HO + 2N$$

Aniline. Nitrous Acid. Carbolic Acid. Water. Nitrogen.

The experiment succeeds best by heating hydro-chloride of aniline with nitrate of silver $=AgO,NO_3$.

It appears possible that hereafter aniline may be advantageously produced by distilling nitrogenous substances, such as bone black, &c., together with the bituminous coal.

Dr. R. Wagner* has experimentally proved recently that, by passing superheated steam (at 300° C.=572° F.) and vapors of phenic acid (heavy tar oils) over alkaline cyanides, for instance cyanide of barium, the amount of aniline is largely increased.

---

* Jahresbericht, 1860; also L. E. Krieger's repeatedly mentioned treatise, p. 45.

Artificial Preparation of Aniline.

A much more fruitful source whence aniline may be prepared is benzole, entering into the composition of coal tar naphtha. Benzole, or the hydride of phenyle, has the formula $= C_{12}H_6$ or $C_{12}H_5H$. The process of manufacturing aniline from benzole consists in the following two operations:—

1. Conversion of benzole into nitro-benzole.
2. Reduction of nitro-benzole into aniline.

The preparation of nitro-benzole on a commercial scale is accomplished in a worm-like apparatus made of glass or stoneware. The upper end terminates like the prongs of a fork in two branches. Through one of these openings flows benzole, and through the other a fine stream of fuming nitric acid, or a mixture of commercial nitric acid with half its volume of sulphuric acid. The worm is surrounded by cold water. The nitro-benzole collected at the lower end of the worm is first washed with water, then with a solution of carbonate of soda, and next deprived of its water by means of chloride of calcium and rectified.

Nitro-benzole has the chemical formula

$$C_{12}H_5NO_4,$$

or, more rationally expressed,

$$C_{12}\begin{cases} H_5 \\ NO_4 \end{cases}$$

Its formation from benzole is explained by the following equation:—

$$\underbrace{C_{12}H_5 + H + NO_5}_{\text{Benzole or hydride of phenyle and nitric acid.}} = \underbrace{C_{12}H_5NO_4 + HO}_{\text{Nitro-benzole and Water.}}$$

Transformation of Nitro-benzole into Aniline.

Nitro-benzole when acted upon by different reducing agents such as sulphide of hydrogen, metallic zinc, acetate of the protoxide of iron, and acetic acid, loses four equivalents of oxygen, and takes up two of hydrogen, being thereby converted into aniline. Béchamp recommends that for the advantageous reduction of one part of nitro-benzole, there be employed one part of acetic acid, and one and one-half parts of iron filings; the reaction may be expressed as follows:—

$$\underbrace{C_{12}H_5NO_4}_{\text{Nitro-benzole.}} + \underbrace{2\overline{Ac}}_{\text{Acetic Acid.}} + \underbrace{4Fe}_{\text{Iron.}} =$$

$$\underbrace{C_{12}H_7N}_{\text{Aniline.}} + \underbrace{2(\overline{Ac}Fe_2O_3)}_{\text{Acetate of Peroxide of Iron.}}$$

The reaction takes place in a retort of iron, and the mixture becoming hot by itself, without the aid of external heat, the vapors are condensed in a well-cooled receiver containing some acetic acid. The condensed products consist of aniline, acetate of aniline, and also some free nitro-benzole. These are retained in the retort, and distilled to dryness.

The distilled liquor is treated with fused caustic potash, whence the aniline separates as an oily layer, which, after being removed, is distilled again.

The residue in the retort still contains a good deal of aniline, which is extracted with sulphuric acid, and the solution filtered and evaporated to dryness. The remainder is sulphate of aniline, from which, by means of an excess of potassa liquor and rectification, the aniline is obtained. Theoretically we ought to obtain from one equivalent of nitro-benzole, one equivalent of aniline, *i. e.*, 75.5 per cent. In practice we obtain about $\frac{2}{3}$, *i. e.*, 50 per cent.

Chemical Properties of Aniline.

Pure aniline is a colorless liquid, having an aromatic odor, and a burning taste. It is slightly soluble in water, and easily soluble in alcohol and ether. The commercial article is generally brown colored, and of a coal oil-like odor.

Chemical Test for Benzole.

As it becomes often necessary to examine a mixture of oils for benzole, and as there are as yet such contradictory statements in regard to its presence in petroleum, the following test by Prof. Hofmann may be applied:—

A drop of the mixture is heated in a test-tube with some fuming nitric acid, to convert the benzole, if present, into nitro-benzole. A quantity of water is then added to precipitate the nitro-benzole in small drops, which must be taken up by ether. The ethereal solution is then poured into another small tube, and equal volumes of alcohol and diluted hydrochloric acid are next added and a few fragments of granulated zinc dropped in. In a few minutes sufficient hydrogen will be disengaged for reducing the nitro-benzole into aniline; the latter is found to be combined with the acid. The liquor is super-saturated with an alkali and shaken with ether, which dissolves the aniline thus set free. A drop of this ethereal solution allowed to evaporate in a watch-glass, and mixed after the evaporation of the ether with a drop of a solution of hypochlorite of lime, will show the violet tints which characterize aniline. The operations may be executed rapidly and easily.

Aniline is now prepared by the ton to satisfy the constantly increasing demands of industry; hence the consumption of benzole has become so great that none can be imported from England, formerly the chief place of export for the European Continent.

The tar of all the gas works, which, at least where the retorts of Chamotte are employed, does not contain over 1 to 1½ per cent. of benzole, is insufficient to meet the demand, and consequently much of it will have to be directly distilled from coal. Fortunately great purity of benzole is not required in the manufacture of the so-called tar colors, if it is at all advantageous.

Thus Hofmann considers that there exists a necessity of mixing aniline and toluidine, to produce aniline red.

Preparation of Aniline Colors.

Nearly all oxidizing substances in contact with aniline produce coloring. Many different receipts are frequently recommended for preparing one and the same color. We have practically convinced ourselves, that by varying the quantitative proportions of the materials, changing the temperature of the bath, and employing different mordants, almost any variety of colors and shades of color may be obtained, at least on wool and silk.* The colors on cotton appear less fine and varied.

Until a comparatively recent period, objections were raised against aniline colors, on account of their want of durability when exposed to sunlight, etc.; it was said that these, equalling in beauty the tints of flowers, had also their fragility, but even these objections have now in a great measure been overcome, and in some cases entirely removed.

The colors are fixed upon fabrics with and without mordants. The following mordants are usually employed: alum and cream tartar, cream tartar and tin composition, tannin, etc. To fasten the dyes upon cotton fabrics, the goods have to be albuminized, or prepared in an oil bath† like turkey red, or in a soap bath.

In printing goods with aniline colors the solutions are thickened with albumen, gluten, gum, etc., and the printed goods steamed.

* A great variety of colored patterns, including, besides the leading colors, also olive, drabs, brown, etc., have been prepared for the museum of the Department of Agriculture.

† Composed of sweet oil, sulphuric acid, alcohol, and water.

In the following we give a few practical receipts to prepare leading colors; to enter more fully into the matter would require a treatise by itself.*

1. *Aniline Red,* syn., *with Fuchsine, Fuchsiacine, Rosaniline, and Magenta.*

It may be obtained in the following manner according to Brooman:—

Three parts of anhydrous bichloride of tin (spiritus libavii fumans) are slowly poured into four parts of aniline, stirring the mixture constantly, and heating it to boiling for fifteen to twenty minutes.

When cold the mass is pasty, and the coloring principle may be extracted from it by boiling with large quantities of water and filtering the solution whilst hot. The fuchsine separates as the liquor grows cold.

---

* To those interested in this new and highly interesting branch of industry, we would recommend the following treatise as the best: "Theorie und practische Anwendung von Anilin in der Färberei und Druckerei, von Ludwig I. Krieg 2te Aufl." Berlin, 1862.

Additions of chloride of sodium, Rochelle salt, etc., deposit it more completely. It is collected and dissolved in hot water, alcohol, or wood-spirit, and the solutions employed for dyeing.

Fuchsine is also soluble in ether, benzine, and bisulphide of carbon.

The fuchsine when evaporated to dryness assumes a metallic, golden green aspect.

In the following table are enumerated some of the other processes by which fuchsine, frequently of a somewhat different composition, may be prepared from aniline.

MANUFACTURER.	REAGENT.	TIME OF EXPOSURE.	TEMPERATURE.
Renard, Franc & Co.	Anhydrous bichloride of tin	15 to 20 minutes	180° C. = 356° F.
Gerber & Keller	Pernitrate of mercury	8 to 9 hours	100° C. = 212° F.
Hofmann	Bichloride of carbon	30 hours	190—200° C. = 374—392° F.
Natanson	Chloride of ethylene (Dutch liquid)		200° C. = 392° F.
Girard & De Laire	Arsenic acid	4 to 5 hours	160° C. = 330.8° F.
Persoz, de Luynes, Salvetat	Arsenic acid	7 hours	180° C. = 356° F.
Depouilly & Lauth	Igniting nitrate of aniline		200° C. = 392° F.
Dale & Caro	Nitrate of lead	1 to 1½ hours.	180—193° C. = 356—379.4° F.
Hughes	Nitric acid		150—200° C. = 302—392° F.

2. *Aniline Violet,* syn., *with Violine, Indisine, Pourpre Française, Anileine, Phenamein, Mauve.*

There exist different shades of this color, either the blue or red tint prevailing, whence the adoption of the additional names—

Aniline purple,

Roseine (color of the rose).

They are evidently most closely allied, are formed under the same circumstances, and have almost identical properties toward chemical reagents.

Purple.

W. Perkins and A. H. Church mix equivalent proportions of sulphate of aniline (Toluidine, &c.) and bichromate of potassa. The black precipitate is filtered off, washed with water until free from sulphate of potash, and dried. The dry product is treated with coal tar naphtha, to extract resinous matter until the solvent ceases to be brown. After this, the mass is repeatedly boiled with alcohol or wood spirit, which extracts the desired coloring principle. The solutions, when distilled in a retort to regain the solvent, leave a beautiful bronze-colored substance behind. It has a reddish hue, and is known as Aniline purple.

Violine.

Dr. Price proceeds thus:—

1 equivalent of aniline=(93 parts).

2 equivalents of sulphuric acid (98 parts spec. grav.=1.850), are mixed.

20 parts of water added and the whole boiled (212° F.).

1 equivalent of finely pulverized binoxide of lead (119.6 parts) is added next, and the whole kept boiling for some time and filtered whilst hot. The filtrate is distilled with caustic potash or soda, until all free aniline has passed over. The residue is thrown on a filter and slightly washed with water, and then dissolved by a dilute boiling solution of tartaric acid. After filtering, the solution may be concentrated and used in dyeing or boiled down to dryness, and the mass dissolved in alcohol, and thence

be obtained by evaporation as a bronze colored solid.

Instead of binoxide of lead, peroxide of manganese may be substituted.

The dye, together with some protoxide of manganese, is thrown down from the filtered alcoholic extract by caustic potash, and the violine redissolved in alcohol.

Prof. Bolley states that a violine bath may be prepared by treating a solution of sulphate of aniline with chlorine water or a weak solution of chloride of lime. Silk assumes a fine violet color in the bath when warmed.

According to the same chemist, aniline red and violet have the same chemical composition and are isomeric modifications.*

The annexed directions prove that a slight change in the proportion of ingredients produces alteration in color.

* L. J. Krieg, p. 141.

Purpurin (Couleur de pourpre).

2 equivalents of aniline (186 parts).
2 equivalents of oil of vitriol (98 parts).
3720 parts of water.
1 equivalent of binoxide of lead (119.6 parts).

Or,

50 parts of aniline.
26 parts of oil of vitriol 66° B.
1000 parts of water.
32 parts of binoxide of lead.

Roseine (Couleur de rose).

1 equivalent of aniline (93 parts).
1 equivalent oil of vitriol (49.0 parts).
1860 parts of water.
2 equivalents of binoxide of lead (239.2 parts).

Or,

50 parts of aniline.
26 parts oil of vitriol 66° B.
100 parts of water.
128 parts of binoxide of lead.

A slight agitation of aniline purple with moist binoxide of lead furnishes Roseine.

Aniline blue, according to A. Schlumberger,* Basel.

1 part of aniline red is mixed with 3 parts of aniline and $1\frac{1}{2}$ part of acetic acid, and the mass treated with sufficient carbonate of caustic soda to decompose the acetate of aniline formed whilst acetate of soda is now being produced. The mixture is heated for some time at a temperature between 356° F. and 410° F., until the desired shade of blue appears. The product is precipitated with strong hydrochloric acid, and heated to boiling, whence the blue dye separates in a solid state, and can be removed from the liquor with a ladle. To get rid of the adhering acid, it is repeatedly boiled with water, pressed and dried. That portion of coloring matter dissolved by the strong acid can be regained upon the addition of water, which precipitates a blue of a second quality, having more of a reddish tint.

The dried blue dye is soluble in alcohol

---

* Lond. Jour. of Arts, and Prof. R. Böttger's Polytechnisches Notizblatt, Jahrgang, xix. 1864.

and wood spirit, and these solutions are used for coloring.

Aniline brown.

G. de Laire, of Paris, has taken out a patent in England for its manufacture. It is obtained by acting upon aniline violet or blue with a salt of aniline, thus:—

To 1 part of violine, add—

4 parts of anhydrous chloride of aniline, raise the temperature of the solution rapidly to 464° F., maintaining it for one to two hours, until the mass turns brown, and yellow vapors with a strong garlic odor are given off.

The brown coloring matter is soluble in water, alcohol, and acids, and may in this form be used for dyeing. Kitchen salt precipitates it from the solution, and serves thus to purify it still further.

Instead of aniline violet, the material producing it, such as arseniate of aniline, can be used.

Aniline green, according to Usebe.*

An aniline salt is dissolved in hydrochloric, sulphuric acid, &c., and common rectified aldehyd $= C_4H_4O$ added and the mixture left standing at a common temperature for eighteen to twenty-four hours, whence it assumes a greenish-blue hue; it is then diluted with water containing a little acid to prevent the blue dye from falling, and gradually hyposulphite of soda added. It has to be seen that the mineral acid used for solution is constantly kept in excess. By heating now to boiling, sulphurous acid is evolved and sulphur thrown down. The solution is filtered whilst hot. If an excess of the hyposulphite was employed, the filtrate is yellowish-green.

The colors produced from the chemically interesting coal-tar hydrocarbon, naphthaline, are as yet of but slight technical signification, and may be omitted by us.

* Schweizer Polys. Zeitschrift, 1864, p. 77.

APPENDIX.

AMOUNT OF PETROLEUM EXPORTED FROM NEW YORK IN 1863 AND 1864, AND THE COUNTRIES AND PLACES TO WHICH IT WAS SENT.

THE increase in the oil trade during the last two years is owing to the increased foreign demand. For home consumption alone, a comparatively small fraction of the figures given below would suffice to glut the market. Europe has greatly exceeded the United States in the multiplicity of uses to which petroleum has been put. Of late discoveries of its properties have so increased in number abroad that the supply proves inadequate to the demand.

The table shows how extensive the demand has already become, and, together with our account of the article itself, will serve in some degree to foreshadow the future. It is impossible to foresee, with even an approximation to correctness, the extent to which this product will become an article of exportation and of usefulness throughout the world.

We republish from the *New York Shipping and Commercial List:*—

	1864.	1863.
To Liverpool	734,755	2,156,851
London	1,430,710	2,576,331
Glasgow, &c.	368,402	414,943
Bristol	29,124	71,912
Falmouth, England	316,402	623,176
Grangemouth, England	......	425,334
Cork, &c.	3,310,362	1,532,257
Bowling, England	87,164	
Havre	2,324,017	1,774,890
Marseilles	1,982,075	1,167,893
Cette	4,800	
Dunkirk	232,803	
Dieppe	79,581	46,000
Rouen	......	143,646
Antwerp	4,149,821	2,692,974
Bremen	971,905	903,004
Amsterdam	77,041	436
Hamburg	1,186,080	1,466,155
Rotterdam	532,926	757.249
Gottenburg	33,813	

Cronstadt	400,376	88,060
Cadiz and Malaga	55,674	33,284
Tarragona and Alicante	16,823	33,000
Barcelona	25,500	
Gibraltar	69,181	308,450
Oporto	17,474	2,339
Palermo	7,983	57,115
Genoa and Leghorn	635,121	399,674
Trieste	165,175	3,000
Alexandria, Egypt	4,000	
Lisbon	167,195	64,662
Canary Islands	3,358	5,125
Madeira	……	400
Bilboa	2,500	
China and East Indies	34,338	36,942
Africa	25,195	12,230
Australia	377,884	304,166
Otago, N. Z.	10,810	5,500
Sidney, N. S. W.	97,880	48,013
Brazil	149,676	160,152
Mexico	112,986	69,481
Cuba	418,034	356,436
Argentine Republic	20,260	24,470
Cisalpine Republic	78,552	117,626
Chili	92,550	66,550
Peru	169,061	256,407
British Honduras	6,072	440
British Guiana	7,881	15,104

British West Indies . .	70,976	60,931
British N. American colonies	28,902	16,995
Danish West Indies . .	8,463	31,503
Dutch West Indies . .	26,638	12,143
French West Indies . .	16,020	9,104
Hayti	7,088	12,064
Central America . .	993	456
Venezuela . . .	28,583	15,455
New Granada . . .	56,490	107,837
Porto Rico . . .	20,026	59,439
Total gallons . .	.21,280,489	19,547,604

During the following years there has been exported from other ports as follows:—

	GALLONS.		
	1864.	1863.	1862.
Boston . . .	1,696,307	2,049,431	1,071,100
Philadelphia .	7,760,148	5,595,738	2,800,978
Baltimore . .	929,971	915,866	174,830
Portland . .	70,762	342,082	120,250
	10,457,188	8,703,117	4,167,158

The total exports from the United States are—

	Gallons.	Value from average prices.
1862	10,887,701	
1863	28,250,712	$14,616,923
1864	31,745,687	23,686,457
No. of barrels of 40 gallons each,		1,772,102½

16*

AVERAGE PRICES OF PETROLEUM IN 1864 AT NEW YORK AND PHILADELPHIA.

	Crude (per gallon).	Refined (per gallon).
January	$31\frac{3}{16}$ cents.	$52\frac{5}{8}$ cents.
February	$30\frac{1}{4}$ "	$55\frac{3}{4}$ "
March	$31\frac{1}{4}$ "	$59\frac{1}{2}$ "
April	$37\frac{3}{16}$ "	$64\frac{7}{16}$ "
May	38 "	$65\frac{1}{2}$ "
June	$44\frac{1}{2}$ "	77 "
July	$52\frac{1}{16}$ "	92 "
August	$52\frac{5}{8}$ "	$87\frac{3}{4}$ "
September	$46\frac{7}{16}$ "	$85\frac{3}{8}$ "
October	$40\frac{5}{8}$ "	$75\frac{3}{4}$ "
November	$45\frac{7}{16}$ "	$86\frac{3}{16}$ "
December	$52\frac{3}{8}$ "	$92\frac{2}{16}$ "
Average for 1864	41.81 "	74.61 "
Average for 1863	28.13 "	51.74 "

INDEX.

A

	PAGE
Alcohol series of hydrocarbons	42
Aluminum	17
American Druggist's Circular quoted	72, 107
Andrews, Prof. E. B., drawings of the anticlinal lines near which oil originates	110, 112, 113
quoted	150
Aniline	71
artificial preparation of	165
blue	179
brown	180
chemical properties of	168
colors, preparation of	170
origin of, in coal tar	163
preparation of, from coal tar	160, 161
red	172
transformation of nitro-benzole into	166
violet	175
Artificial products, comparison of, with those found in nature	66
Avery well	140

B

Baird, H. C., quoted	15
Barlow, Peter, quoted	15

	PAGE
Barrels	106, 116
Beech-wood tar, ingredients of	31
Benzine, absence of, in American petroleum	133
or naphtha	21
Benzole, absence of, in some American oils	82
chemical test for	168
series of hydrocarbons	42
Bitumen, elastic	68
Bitumens, localities of	90
Bituminization, table showing the decrease in the proportion of oxygen in	132
Bituminous shale, theory of derivation of petroleum from	134
oil, history of	62
slate, distillation of	21
yield of, on distillation	52
Black shales, oil in	139
Bogs	129, 130
Boring, new process of	119
of oil wells	109
time required for	117
Brawley well	140

C

Cailletet, experiment of	131
Canada, oil in	99, 100
oils	80
qualities of	101
Canadian oil region, dip of rocks in	141
Cannel coal, products of the distillation of	40
Caoutchouc, mineral	68
Chemical composition of petroleum	70, 73
Chemistry, influence of, on the mechanic arts	13

INDEX.

	PAGE
Chrysene	43
Coal as a fuel, national importance of	14
beds, geographical connection of oil with	128
under Rangoon oil wells	129
distillation of	21
from the Tyne	133
in the United States	14
measures, origin of	129
oil in	139
Pennsylvania anthracite	133
semi-anthracite	133
South Staffordshire	133
the power exerted by	15
wood or turf, products obtained from	24
Coal oil or kerosene, purification of	60
tar creasote	35
Coal tar, distillation of	45
Comparison of artificial products with those found in nature	66
Cornwall well	140
Creasote	21, 32

D

	PAGE
De La Rue & Miller, examination of Birmese naphtha	82
Derricks, erection of	111
Deville, experiments of	131
Devonian rocks, oil found in Canada and New York	136
Distillation, dry, of organic bodies	24
of coal tar	45
Drill	111, 120
Drill-stem	120
Dry distillation, petroleum must have been produced by	133
or destructive distillation of organic bodies	24

E

	PAGE
Elastic bitumen	68
Empire spring	98
Empyreumatic oil, history of	62
Engine, derrick, and connections	118
Engines	114
advantages of, over spring poles	114
Eupion or light oil	31
Export of petroleum	181

F

Faraday, the manner in which he originally obtained benzole	45
Fires in wells	103
Frankland, Dr., quoted	87
French Creek	97
French process of boring	119
Fuchsine	172
processes for manufacture of	173

G

Geological distribution of petroleum in the United States and Canada	155
Gesner, Dr., quoted	93

H

Hall, Prof. James, quoted	144
Hard rocks contain small quantity of petroleum	109
Hatchetin	69
Heat, source of	131
Heavy oil	32
Heer, Oswald, quoted	138
History of bituminous and kerosene or empyreumatic oils	62

	PAGE
History of petroleum or rock oil	89
Hitchcock, Professor, quoted	14
Hofmann, Prof., quoted	168
Humfrey, Charles, quoted	75
Hydrocarbons	42

I

Idrialin	69
Illuminating power of petroleum	87

J

Jarr	121

K

Kapnomore	32
Kerosene oil, history of	62
purification of	60
Kildare, chemical works at	53

L

Lake tar	93
Lava, marsh gas from	131
Lebon, his application of carbo-hydrogen gases to illumination	27
Lesley, Prof. J. P., quoted	139
Lesquereux quoted	138
Light oil, or eupione	31
Lignite	133
Lignites, tertiary, in Trinidad, Lombardy, and Middle Asia	129
Limestones impregnated by petroleum	109
Locating wells, hints in	149
Logan, Sir William, quoted	141
Lyell, Sir Chas., quoted	135

	PAGE
Magenta	172
Magnesium	17
Mahoning oil region, dip of rocks at	140
Manufacture of photogenic oils, &c., from coal, wood, and turf	44
Manfield & Young's patents	39
Marsh gas from lava	131
given off by plants in peat bogs	129
Mineral caoutchouc	68
Murdoch, his application of coal gas to illumination	27
Muspratt, Dr. S., analysis of Canadian petroleum	79

N

Naphtha or benzine	21
Naphthaline	42
and paranaphthaline, formation of	44
Natural science, influence of	13
Nature, hidden force of	13
New York, oil in	99
Nitro-benzole, transformation of, into aniline	166
Ohio, oil in	98
oil springs in	110
Oil and coal regions, apparent connection of	139
business, magnitude of	106
Creek	94, 95, 96
geological position of oil in	105
rocks, dip of	139
first obtained by boring, 1819	96
formation, theories of	126, 127
may have been forced by pressure far away from place of original distillation	128
photogenic	24
tools, where to be had cheapest and best	117
wells, cost of sinking	111, 114

	PAGE
Oil wells, hints in locating	149
Origin of petroleum	124
Organic bodies, dry or destructive distillation of	24
origin of petroleum	125
Ozokerite, or fossil wax	69

P

Paper, raw materials for manufacture of	19
Paraffine, or tar-wax	21, 37, 42
series of hydrocarbons	42
Paranaphthaline	43
Parrish, Edw., quoted	103
Peat	133
bogs, plants in, give off marsh gas, &c.	129
or turf from Hanover, yield of	52
Pelouze & Cahours	81
on American petroleum	133
Pennsylvania, oil in	96, 97
oils, qualities of	100
petroleum in	108
Percy, Dr., experiments of	132
Petroleum, export of	181
discovery and development of, properties of	20
geological distribution of	185
position of	109
history of	89
illuminating power of	87
impregnates limestone, sandstone, and shales	109
in Pennsylvania	108
native, discovery of	22
on the American Continent	76
origin of	124
or rock oil, its chemical composition	70
prices of	185

17

	PAGE
Petroleum, qualities of those from different localities	101
refining of	84
Pettenkofer's apparatus for manufacture of gas from wood	27
Petzholdt, Dr. A., experiments	126
Phillips well	98, 140
Photogenic oils	24
manufacture of, from coal, wood, and turf	44
Picamar	32
Pilla, observation of	131
Pipe-tongs	121
Pittsburg petroleum	108
Preparation of aniline colors	170
Pressure, effect of, in forcing condensed volatile products to seek a high level	128
Prices of petroleum	185
Products of distillation of cannel coal and their chemical composition	40
of wood, coal, or turf	26
Purification of coal oil	60
Purple	175
Purpurin	178
Pyrene	43

Q

| Quinn, A., patent still | 85 |

R

Rangoon oil, analysis of	75
Reamer	114, 120
Refining of petroleum	84
Reichenbach, discoveries of	31, 39
discoveries of constituents now prepared from petroleum	20

	PAGE
Retinite or retinasphalt	68
Retinasphalt or retinite	68
Rocks, dip of, in the Oil Creek region	140
Rogers, Prof. H. D., observations of, on volatile substances in Appalachian coal fields	135
quoted	15
Rope tools, advantages of	116
Roseine	178
Round-reamer	120

S

Sand-pump	114, 121
Sandstones impregnated by petroleum	109
Schorlemmer's analysis of cannel coal oils	134
his examination of oils	82
Scientific American quoted	155
discoveries, facts in regard to	13
important	16
Seed bag	116
Seneca Indians	94
Shales impregnated by petroleum	109
Shaw, Thomas, quoted	71
Spring poles	114
advantages of, over engines	115
Strata at Bristol, Ontario County, N. Y.	145, 146

T

Tanks	116
Tar, beech-wood, ingredients of	31
constituents of, from bituminous slate	52
of, from peat or turf	53
Tar-wax or paraffine	37
Tate, A. N., analyses of petroleum	76
quoted	88

	PAGE
Temperature, variation of results with	44, 55
Temper screw	120
Thermometer for determining the temperatures at which different oils explode	103
Titusville	97
dip of rocks at	140
Troost, experiment of	131
Turf, products obtained from	24

U

Uncertainty of wells	152

V

Vegetable matter, dry or destructive distillation of	25
Venango County	97
Violine	176
Vohl, Prof., analysis of Rangoon oil	75

W

Websky, Dr. Justus, quoted	129
Wells, uncertainty of	152
West Virginia, oil in	98
Wood, coal, or turf, products obtained from	24
Wood tissue	133

www.ingramcontent.com/pod-product-compliance
Lightning Source LLC
Chambersburg PA
CBHW020240170426
43202CB00008B/158